# 2022 NIGHT SKY ALMANAC

A Month-by-Month Guide to North America's Skies from The Royal Astronomical Society of Canada

Nicole Mortillaro

FIREFLY BOOKS

A FIREFLY BOOK

Published by Firefly Books Ltd. 2021


First printing

**Library of Congress Control Number: 2021936331**

**Library and Archives Canada Cataloguing in Publication**
Title: 2022 night sky almanac : a month-by-month guide to North America's skies from the Royal Astronomical Society of Canada / Nicole Mortillaro.
Other titles: Two thousand twenty-two night sky almanac | Twenty twenty-two night sky almanac
Names: Mortillaro, Nicole, 1972- author. | Royal Astronomical Society of Canada, issuing body.
Description: Includes bibliographical references.
Identifiers: Canadiana 20210186186 | ISBN 9780228103264 (softcover)
Subjects: LCSH: Astronomy—Canada—Observers' manuals. | LCSH: Astronomy—Canada—Amateurs' manuals. | LCSH: Astronomy—United States—Observers' manuals. | LCSH: Astronomy—United States—Amateurs' manuals. | LCSH: Astronomy—Popular works. | LCGFT: Handbooks and manuals.
Classification: LCC QB64 .M672 2021 | DDC 523—dc23

Published in Canada by
Firefly Books Ltd.
50 Staples Avenue, Unit 1
Richmond Hill, Ontario L4B 0A7

Published in the United States by
Firefly Books (U.S.) Inc.
P.O. Box 1338, Ellicott Station
Buffalo, New York 14205

Project manager/Editor (Firefly Books): Julie Takasaki
Project manager (RASC): Robyn Foret, President of The RASC
Copyeditor: Elizabeth Howell and Nancy Foran
Technical editors: David M.F. Chapman FRASC (Emeritus Editor, RASC *Observer's Handbook*) and James S. Edgar FRASC (Editor, RASC *Observer's Handbook)*
Graphic design: Noor Majeed
Illustrations and sky charts: Peter Kovalik
Printed in China

Canada

We acknowledge the financial support of the Government of Canada.

# Contents

# Introduction

If you're reading this, you clearly love the night sky and all the joy it brings you.

This guide aims to provide novice and intermediate amateur astronomers with the knowledge they need to enjoy the night sky in 2022, from understanding planets, nebulae, comets, asteroids and meteors, to learning about annual and significant celestial events. It's your pocket guide to the cosmos.

But first, it's important that you understand how we navigate the night sky. Just like for navigation on Earth, astronomers use particular coordinates in the sky to figure out what we're looking at.

The **celestial sphere** is an imaginary sphere with Earth at its center. At any one time, an observer's night-sky view only includes half of this sphere, because the other half is below the horizon.

Earth's axis is tilted at 23.5 degrees to the **plane** of the Solar System – the plane being the orbit of the Earth around the Sun. For observers on the ground, the celestial sphere seems to rotate from east to west; that is why the Sun, for example, rises in the east and sets in the west.

The Earth's axis points less than 1 degree away from **Polaris**, the North Star, in the Northern Hemisphere. It's really important to know where north is, so you can navigate around the sky. Use a star chart, your smartphone or both to find true north at your observing location. (In the south, the Earth's axis points about one

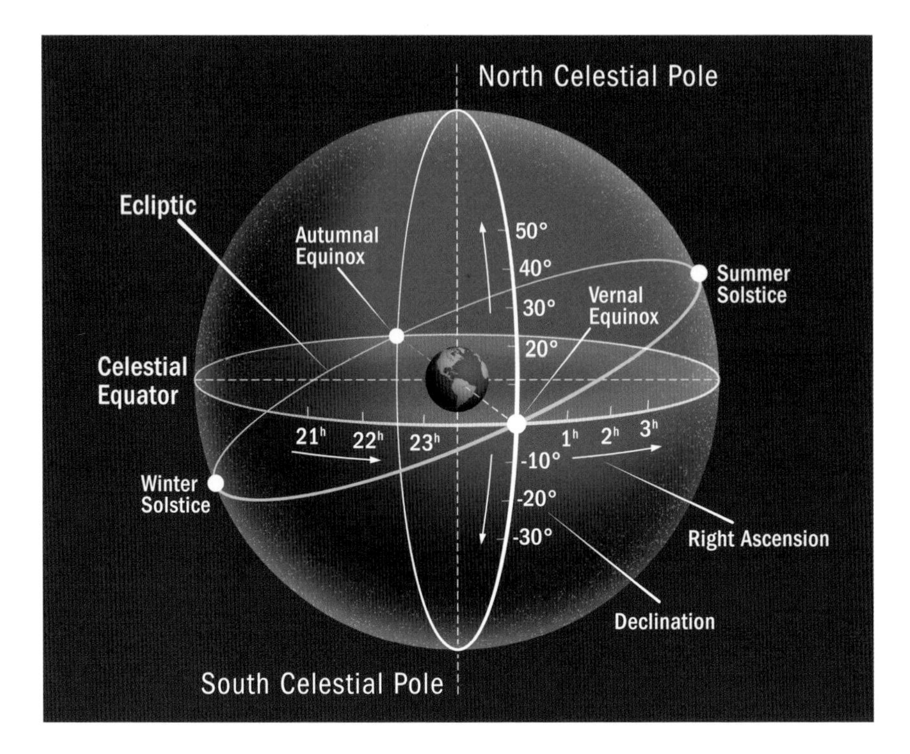

degree away from **Sigma Octantis**, a faint star; given Canada and the United States are in the Northern Hemisphere, however, we will focus on Polaris). When you look at the sky over long periods of time, Polaris stays still as the other stars rotate around it.

You can easily see this phenomenon by setting up a camera, pointing it toward Polaris, and taking an extended (or long-exposure) photograph of the stars. You will see circular streaks in the sky as other stars rotate around Polaris. Some stars are so close to the North Celestial Pole that they never set. Other stars farther from the pole will rise in the east and set in the west, just like our Sun. Many stars are so far from the pole that they never rise above your horizon.

While beginner astronomers will navigate using a simplified star chart, it's helpful to know some of the terminology we use for finding our way around the sky. That way, as you gain confidence, you can follow professional astronomers in their observations.

The point directly overhead your observing location is called the **zenith**, and the **celestial equator** is directly above the Earth's equator. The point directly below you (under the horizon and opposite to the zenith) is the **nadir**. The line that runs from the north point on the horizon, through the zenith, to the south point on the horizon, is the **meridian**.

To navigate the night sky, astronomers created a type of latitude and longitude called **celestial coordinates**. In astronomy, we use **declination** (Dec.) and **right ascension** (RA). Declination is like latitude on Earth, running from north to south. Right ascension is like longitude on Earth, running from east to west. Declination is measured in degrees, and right ascension is measured in hours, minutes and seconds.

Star trails with Polaris at the center

As Earth rotates, the apparent position of most of the stars changes. More advanced astronomers may want to learn more about exactly where to find faint stars in their telescopes. For these situations, we navigate using **altitude** (angular elevation above the horizon, between 0 degrees at the horizon and 90 degrees at the zenith) and **azimuth** (the number of degrees clockwise from due north). But if you're just starting out, don't worry yet about these advanced observing techniques. A simple star chart will help you find the constellations and the planets.

If you want to look for planets, you should also pay attention to the **ecliptic**. That is the path the Sun takes through the constellations, and the Moon and planets do not stray far from that path.

# Handy Sky Measures

Astronomers need to know how far apart things are in the sky. And they do this using angular degrees.

It's not hard to imagine the sky as a sphere that measures 360 degrees – after all, space surrounds Earth on all sides. Standing in one spot on Earth, if you trace the sky from horizon to horizon, that would equal 180 degrees. Remember, the other 180 degrees is under the horizon.

If you want to measure the distances between two objects – say, between the Moon and Venus as they appear together in the sky – you can use your hand as a measuring tool. It all lies within your fingers.

Hold your hand at arm's length. The width of your pinky finger equals 1 degree. The width of your three middle fingers held at arm's length equals five degrees; a closed fist is 10 degrees; the distance between the tip of your index finger and the tip of your pinky is 15 degrees; and the distance between your thumb and pinky is roughly 25 degrees.

This is particularly helpful when trying to see how high or low something is above the horizon.

You can practice measuring the degrees with stars found in the Big Dipper. The chart shows how to find the Big Dipper using Polaris.

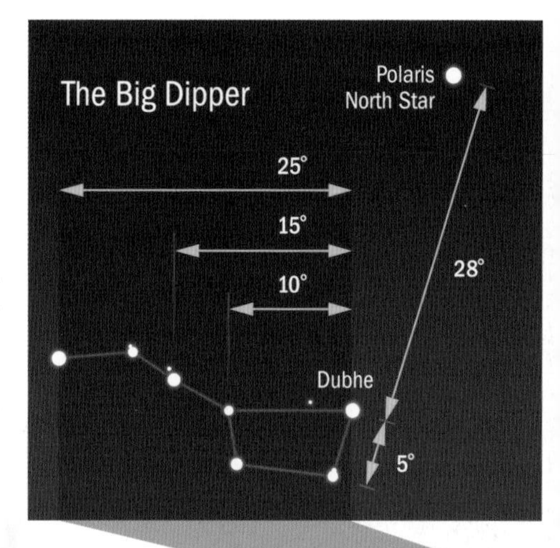


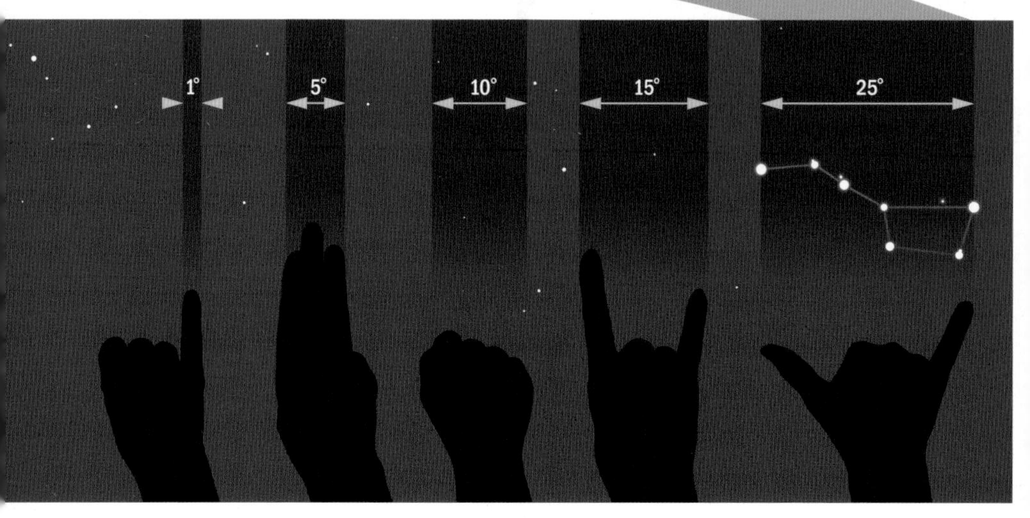

A view of Polaris and the Big Dipper
over Dinosaur Provincial Park, Alberta

# Binoculars and Telescopes

Using your eyes to move around the sky allows you to see the Moon, the planets, the Sun and even some objects outside of our Solar System. However, having binoculars or a telescope opens up the Universe to you.

Binoculars are an amazing first tool for seeking out the marvelous wonders of the night sky. A good pair of binoculars can reveal the intricacies of the Moon, showing its peaks and valleys, or uncover the daily motion of Jupiter's moons as they dance around our Solar System's largest planet. A decent telescope can make these objects appear bigger and show details of star clusters or nebulae (gas clouds in space).

But the question that often arises is what type of binoculars or telescope should I get?

For binoculars, it's important to understand two things: the **magnification** (or power) and the **aperture**. Binoculars are usually represented by two numbers, separated by an "×." Two example binocular numbers are 7×50 or 10×50. The magnification is the first number, and it represents the number of times larger something will appear compared to viewing it with the naked eye. The second number is the aperture in millimeters (fractions of an inch), and it represents the diameter of each lens. Good viewing in part depends on how well your binoculars can gather light. If your binoculars have a larger second number, they have a larger aperture and can gather more light from distant objects. But that comes with a cost. Bigger binoculars are heavier and, therefore, more difficult to hold, which means you will need a tripod to steady your view.

Another number often given for binoculars is the **field of view**, or FOV. This is how wide you will be able to see in degrees (see "Handy Sky Measures," on page 6). In general, the higher the magnification you use, the smaller your field of view will be. At times, this will mean you need to make a choice. Do you want to zoom in on a particular crater on the Moon, for example, or observe a more spread-out chain of mountains? Making these decisions is difficult for astronomers, too, so don't worry if it feels hard – it gets a bit easier to decide what to do with practice.

The top end of handheld binoculars is typically 7×50. Larger than that, it's best to get a tall tripod if you want the best view possible and to share views with others.

We recommend using binoculars for a few months before investing in a telescope. Once you are comfortable with binoculars, luckily, your knowledge will be useful because telescopes use many of the same definitions for observing.

Remember that light-gathering ability is important; a typical beginner's telescope usually ranges between 50 to 150 millimeters (2 to 6 inches) in aperture, the diameter of the main lens or mirror. The bigger the **aperture**, the more light the lens will gather, and the brighter the view will be. Invest in a sturdy tripod, too, so that the telescope will not shake when you use it.

The **magnification** of a telescope depends on the eyepiece used; magnification is calculated by dividing the focal length of the telescope by the focal length of the eyepiece (using like units). For example, a telescope with a focal length of 1,000 millimeters and a 10-millimeter eyepiece will magnify an object 100 times. The same 10-millimeter eyepiece in a 500-millimeter telescope would magnify an object 50 times.

While a bigger aperture and higher magnification may seem like the best way to go, it's best not to get too caught up in choosing a telescope with the highest numbers. What is important is how you plan to use it. Since most people are unlikely to have backyard observatories, the most important thing may be portability. Too heavy a telescope means you're unlikely to haul it out to the backyard or to a vacation spot, far from city lights.

When it comes to telescopes, there is a wide array of choices. Some of the most popular are refractors, reflectors and compound telescopes. Each type has its own benefits, and it all depends on what you prefer. We therefore recommend you try testing out different telescopes at a local star party held by astronomy groups before making the investment. Alternatively, check reputable astronomy magazines or forums for recommendations for beginners; in most cases, all it takes is a little Internet searching and some patience.

A **refractor telescope** has a front lens that focuses light to form an image at the back, and an eyepiece that acts like a magnifying glass to allow your eye to focus on that image. Refractors tend to be more reliable, as their lenses are fixed in place and, therefore, don't get out of alignment as easily as some other types of telescopes. Refractors are best used for planetary and lunar observing as well as viewing double stars.

A **reflector telescope** uses a mirror to gather and focus light. The advantage for reflectors is they don't usually suffer from **chromatic aberrations**, when light of different wavelengths (i.e., colors) doesn't focus on the same point; this makes them ideal to observe distant objects like star clusters. Chromatic aberrations cause the different colors of the spectrum to split and the image to appear blurry, which isn't great when looking at a group of stars. A **Dobsonian telescope**, which is a variant of a reflector with a simple mount, gives you a far bigger aperture for far less the cost of other telescopes.

**Compound** or **catadioptric telescopes** combine lenses and mirrors to form an image. The **Schmidt-Cassegrain**, a type of compound telescope, has a compact design that makes it quite popular. With this instrument, an astronomer can get a bigger aperture in a smaller sized telescope. This telescope type is also portable, making it easier to move the equipment into remote areas.

Star parties are great opportunities to check out different telescopes and chat with fellow enthusiasts

# Stars

Stars come in many different varieties. Our Sun is a **yellow dwarf star**, on the **main sequence**, meaning that it's converting hydrogen into helium at its core, like most other stars. When a star does this conversion, it releases a tremendous amount of energy. This energy, in the form of sunlight, allows life to thrive on Earth.

The Sun, at 4.5 billion years old, is considered middle-aged. When it dies, five billion years from now, it will at first swell, becoming a **red giant** and engulfing the inner planets, then it will slough off its outer layers to become a **white dwarf**.

One of the most interesting types of stars, some might argue, are **red supergiants**. These colossal stars – roughly 1,400 times the mass of the Sun – have relatively short lifespans.

And when supergiants do die, they do so in a spectacular fashion, in an explosion called a **supernova**. A supernova occurs when a star can no longer convert hydrogen into helium; eventually the core is converted into iron. During a supernova, the star first collapses and then explodes outward, creating even heavier elements.

**Betelgeuse**, a star found in the left shoulder of Orion, is a red supergiant. While many far-away and faint supernovae have been witnessed in modern history in other galaxies, none have occurred in our own galaxy – the **Milky Way**. When Betelgeuse goes supernova, its brightness will rival the full Moon in our sky. If you're hoping to see Betelgeuse explode, you're not alone. However, estimates peg Betelgeuse's death for some time within the next 100,000 years.

The most common star in the Universe is a **red dwarf**, which is a cool star that's much smaller than the Sun. Astronomers often search for potentially habitable planets around these stars, because the inherent dimness and smaller size of red dwarfs makes it easier to spot planets. The stars, however, can be quite volatile, occasionally releasing a tremendous amount of radiation. Intense radiation is not a friendly process for most life forms.

The **Hertzsprung-Russell diagram** (H-R diagram) was developed in the early 1900s by Ejnar Hertzsprung and Henry Norris Russell, following the research findings by two Harvard computers, Annie Jump Cannon and Antonia Maury. The diagram plots the temperature of stars against their luminosity. Just as we do, stars go through certain stages in their lives.

| Top 10 Brightest Stars in the Night Sky | |
|---|---|
| 1. | Sirius |
| 2. | Canopus |
| 3. | Alpha Centauri |
| 4. | Arcturus |
| 5. | Vega |
| 6. | Capella |
| 7. | Rigel |
| 8. | Procyon |
| 9. | Achernar |
| 10. | Betelgeuse |

## The Harvard Observatory Computers

In 1881, Edward Charles Pickering, the director of the Harvard Observatory, hired a team of women to compute and catalog photographs of the night sky. Their work was incredibly important in providing the foundations of astronomical theory. Annie Jump Cannon, one of the computers, added to work done by fellow computer Antonia Maury and developed a system of classifying stars that is still used today.

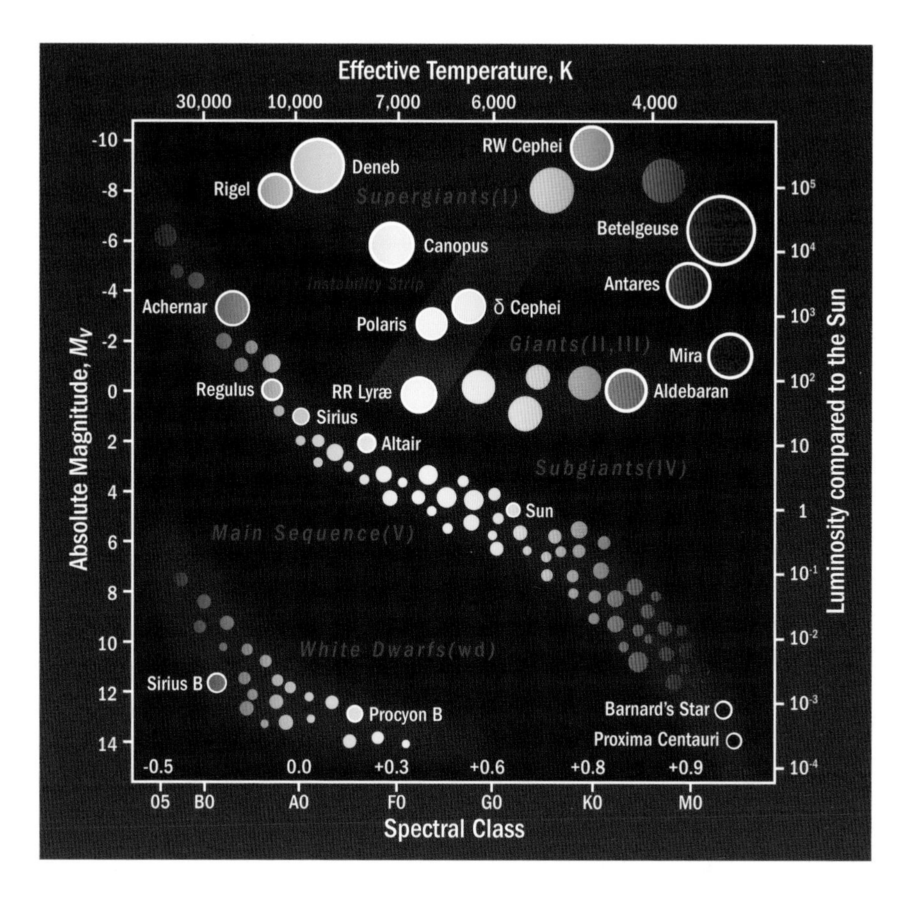


The H-R diagram provides astronomers with the information about a star's current age. Main-sequence stars that are fusing hydrogen into helium – such as our Sun – lie on the diagonal branch of the diagram.

Finally, there's a star's **magnitude**, or apparent brightness. Magnitude is measured on a scale where the higher the number, the fainter it is – and negative numbers are brighter than positive numbers. For example, Sirius, the brightest star in the night sky, measures -1.4 on this scale. Polaris is 2.0, and the Sun is -27. We also use magnitude to measure the brightness of other celestial objects, like the Moon, planets, asteroids and comets.

Note there is a difference between apparent and absolute magnitude. Apparent magnitude is the brightness of an object that we observe from Earth, but absolute magnitude is the brightness of an object if it were placed 32.6 light-years from Earth. This second measure helps astronomers directly compare the magnitudes of objects and is what is used for the H-R diagram. On this scale, Sirius has a magnitude of 1.4, Polaris is -3.6 and the Sun is 4.8.

In astronomy, Greek letters of the alphabet are used to identify stars within a constellation usually from brighter to dimmer.

# Constellations

**Constellations**, a group of stars that make an imaginary image in the night sky, have been around since ancient times. Typically, the images astronomers use are based on Greek, Roman and Arabic mythologies. The International Astronomical Union (IAU) recognizes a total of 88 official constellations.

Some of the most recognizable constellations in the Northern Hemisphere are Orion (the Hunter), Cygnus (the Swan), Leo (the Lion), Gemini (the Twins), Scorpius (the Scorpion) and Ursa Major (the Great Bear).

More recently, there has been more effort to acknowledge constellations of Indigenous peoples. The naming of Indigenous constellations varies around the world, making these constellations regionally distinct. Some groups see Ursa Major as a bear, or a caribou. The Cree see Corona Borealis as seven birds, and Cepheus as a turtle. To the Navajo, Polaris is Nahookos Bikq, meaning central fire. For some Indigenous peoples, such as the Inuit, the Northern Lights are dancing spirits.

Aside from the constellations, there are also **asterisms**, a group of stars within a constellation (or sometimes from several different constellations) that forms its own distinct pattern. The Big Dipper is probably the most famous asterism, as its stars lie within the constellation of Ursa Major. There's also the Summer Triangle, with bright stars from Cygnus, Lyra (the Lyre) and Aquila (the Eagle), and the Winter Triangle with stars from Orion, Canis Major (the Great Dog) and Canis Minor (the Little Dog).

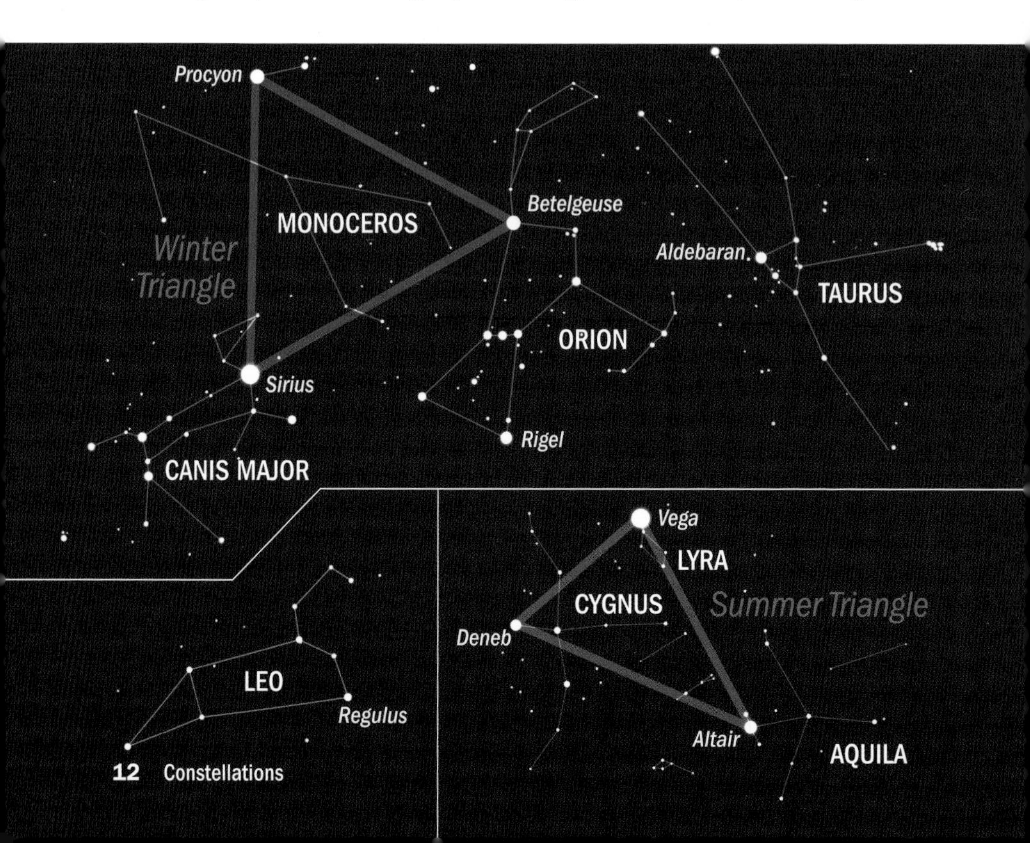

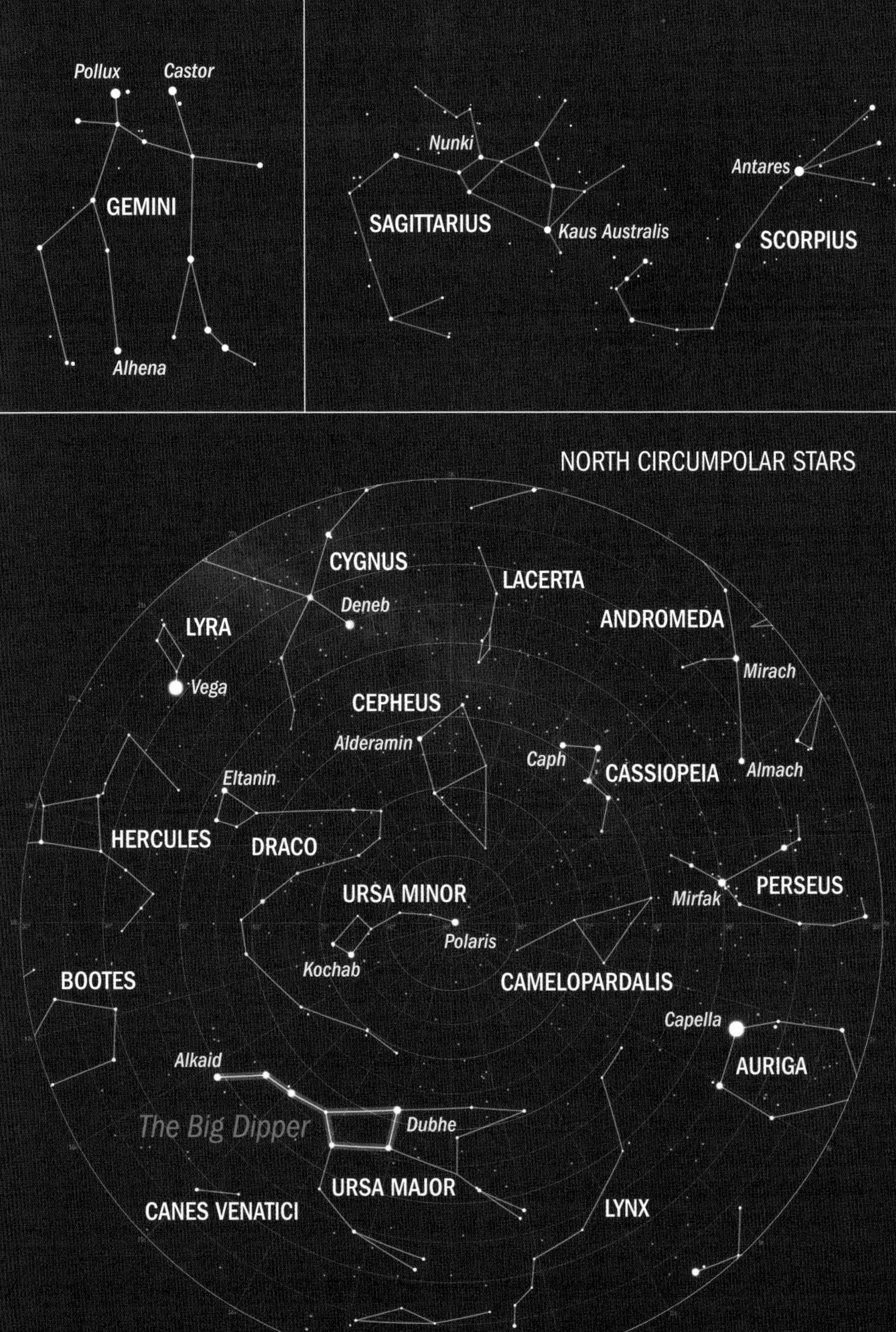
Pollux
Castor
GEMINI
Alhena
Nunki
SAGITTARIUS
Kaus Australis
Antares
SCORPIUS
NORTH CIRCUMPOLAR STARS
CYGNUS
LACERTA
Deneb
ANDROMEDA
LYRA
Mirach
Vega
CEPHEUS
Alderamin
Caph
CASSIOPEIA
Almach
Eltanin
HERCULES
DRACO
PERSEUS
URSA MINOR
Mirfak
Polaris
Kochab
BOOTES
CAMELOPARDALIS
Capella
AURIGA
Alkaid
The Big Dipper
Dubhe
URSA MAJOR
CANES VENATICI
LYNX

# Comets and Meteors

**Comets** are icy balls of debris – specifically, dust and ice – left over from the formation of our Solar System. They are sometimes referred to as "dirty snowballs" and can be stunning objects to see in the night sky when tails of dust and ionized gas fan out behind their cores. While we know of many comets, predicting the appearance of a bright one is all but impossible. Comets tend to fall apart before becoming too bright. The Sun's gravity and heat are strong, and comets themselves are very delicate visitors from the outer Solar System or beyond. Many comets literally crumble under the pressure as they dive in toward the Sun and the inner Solar System, where our planet resides.

Bright comets can be marvelous. On August 17, 2014, Comet Lovejoy C/2014 Q2 was spotted as it came toward the inner Solar System. This comet produced a spectacular tail as it neared the Sun. Tails occur when ice **sublimates**, turning directly from a solid into a gas.

There have been other wonderful naked-eye comets that have graced our night sky, including Comet Hyakutake C/1996 B2, Hale-Bopp C/1995 01, or Comet NEOWISE C/2020 F3. Each comet has the year of its discovery in its official name, so NEOWISE was found in 2020, Hyakutake in 1996, and so on. Many Northern Hemisphere observers were treated to a special show when Comet NEOWISE appeared in the skies in 2020. It was one of the brightest comets to appear in a generation – even city dwellers could spot it through light pollution.

Periodic comets, or ones that we can predict, are given the designation "P," while those that appear unexpectedly are given the designation "C." For example, Halley's Comet, or 1P/Halley, appears roughly every 76 years – a clear P. The last time it passed was 1986; the next time will be in 2061. In general, C comets are brighter than P comets because P comets have sublimated their material into space from repeated trips by the Sun.

A **meteor** (from the Greek *meteoros*, meaning "high in the air") is the light, heat and (occasionally) sound phenomena produced when a

Comet NEOWISE C/2020 F3

The Geminid meteor shower

**meteoroid**, or small debris left in space from comets or **asteroids** (space rocks), collides with molecules in Earth's upper atmosphere. We also call meteors **shooting stars**.

When a meteoroid enters Earth's atmosphere, the surface of the object is heated. Then, at a height typically between 119.8 and 79.9 kilometers (74.5 miles and 49.7 miles), the meteoroid begins to **ablate**, or lose mass. Meteoroid ablation usually happens through vaporization, although some melting and breaking apart can also occur.

Meteoroids are usually the size of a small pebble, but they range in size considerably; some can be as small as the size of the tip of a ballpoint pen, while more unusually, they can be several meters or kilometers across. If a meteoroid is large enough that it doesn't burn up entirely and reaches the ground, it is called a **meteorite** – and we have recorded instances of meteorites causing damage on Earth.

NASA keeps a sharp eye out for threatening objects in space and, so far, has found nothing of concern. We know a big meteorite strike could happen eventually, though. Just ask the dinosaurs, which were likely wiped out by a meteorite at least 11 kilometers (7 miles) across about 66 million years ago. Thankfully, such events are quite rare, happening every few million years on average.

Meteoroids can be divided into two groups: **stream** and **sporadic meteoroids**. Stream meteoroids have orbits around the Sun that often can be linked to a parent object – in most cases a comet, but it can sometimes be an asteroid. When Earth travels within the orbit of a stream, we get a **meteor shower**. Because all the objects (meteoroids) move in the same direction, they all seem to come from one point in the sky, called the **radiant**. The constellation containing the radiant (or in some cases a nearby star) is what gives a meteor its name, such as the Perseids or Geminids (two of the most active meteor showers).

Sporadic meteors just occur in random parts of the sky, with no radiant.

## Meteor Showers in 2022

There is nothing more wonderful than stepping out and looking up at the night sky, only to see a brief streak of light flash against the stars.

Almost monthly, we get major meteor showers. Some showers are best seen from the Northern Hemisphere, while some are better visible in the south, depending on the radiant.

When astronomers talk about the "peak" of a meteor shower, they're referring to the **Zenithal Hourly Rate**, or the ZHR. This is the rate of meteors a shower would produce per hour under clear, dark skies and with the radiant at the zenith. In practice, a single observer will likely see considerably fewer meteors than the ZHR suggests, owing to the presence of moonlight, light pollution, the location of the radiant, poor night vision and inattentiveness.

Here is a list of major meteor showers that you can enjoy simply by staying warm and looking up. No special equipment is needed beyond your eyes; just make sure to give yourself about 20 minutes to adjust to the darkness before searching for meteors.

### Quadrantids: December 27, 2021–January 10, 2022

The Quadrantids might be one of the best meteor showers of the year, with a ZHR of 120 for a brief interval of time. The only thing holding it back from earning the title is that January tends to be cloudy over North America, and the shower's peak has a brief window of six hours, making the average hourly rate closer to 25. But there's good news: the meteors often produce bright **fireballs**, or bright meteors with a magnitude higher than -4.0. This shower's radiant lies between Boötes and Draco. More good news is that there will be no Moon to interfere with your viewing on the peak night.

**Parent Object: Asteroid 2003 EH1**
**2022 Peak Night: January 3–4**

### Lyrids: April 16–30, 2022

After a dearth of meteor showers in the months of February and March, April brings us the Lyrids. This shower produces a ZHR of 18, so it's not a particularly strong shower, but the meteors that do appear tend to produce fireballs. The waning gibbous Moon won't interfere with observations on the peak night, as it rises in the early hours of the morning.

**Parent Object: Comet C/1861 G1 (Thatcher)**
**2022 Peak Night: April 22–23**

### Eta Aquariids: April 19–May 28, 2022

The Eta Aquariids aren't a stellar show for the Northern Hemisphere, since the radiant rises in early dawn, but they can still produce a ZHR of 20 meteors, if you care to get up early enough. Rather than fireballs, this shower tends to produce long trains through the sky. A waxing crescent Moon will cause minimal interference during the peak.

**Parent Object: Comet 1P/Halley**
**2022 Peak Night: May 4–5**

### Perseids: July 17–August 26, 2022

For those in the Northern Hemisphere, the Perseids are considered the best show of the year. The weather is warm and there are fewer clouds at this time of year. The shower's ZHR is 100, though rates of 50 to 75 are more commonly seen on the peak night. Unfortunately, an almost-full Moon will be in the sky this year, so only the brightest meteors will be visible.

**Parent Object: Comet 109P/Swift-Tuttle**
**2022 Peak Night: August 12–13**

### Orionids: October 2–November 7, 2022

This shower is considered medium strength, though it can sometimes surprise us with more activity. On average, the Orionids produce a ZHR of 10 to 20 meteors, though from 2006 to 2009, the shower produced roughly 50 to

75 an hour. The meteors that enter our atmosphere are fast – roughly 66 kilometers (41 miles) per second – and, as a result, produce long, glowing trains. Fireballs are also possible. There will be no Moon to interfere with viewing on the peak night.
**Parent Object: Comet 1P/Halley**
**2022 Peak Night: October 21–22**

**Southern Taurids: September 10–November 20, 2022**

The Southern Taurids last for two months and have several minor peaks. Though this shower produces a ZHR of only 5, it can produce some fireballs. The shower is stronger in the Southern Hemisphere, though we can also catch a few meteors in the north. Unfortunately, a waxing gibbous Moon will be up most of the peak night, which will make it difficult to see any but the brightest meteors.
**Parent Object: Comet 2P/Encke**
**2022 Peak Night: November 5–6**

**Northern Taurids: October 20–December 10, 2022**

Like the Southern Taurids, the Northern Taurids last roughly two months and have a ZHR of 5. There are often reports of an increase of fireballs during the period when the two showers overlap. An almost-full Moon will be in the sky on the peak night, making it challenging to see any but the brightest meteors.
**Parent Object: Comet 2P/Encke**
**2022 Peak Night: November 11–12**

**Leonids: November 6–30, 2022**

While the Leonids don't produce a high number of meteors an hour (they have a ZHR of 15), they can produce outbursts, or meteor storms; the most recent such outbursts occurred in 1999 and 2001. Unfortunately, the next outburst isn't expected until 2099. Still, the Leonids do put on a show, with bright meteors and long-lasting trains. A waning Moon rises after midnight local time on the peak night.
**Parent Object: Comet 55P/Tempel-Tuttle**
**2022 Peak Night: November 17–18**

**Geminids: December 4–17, 2022**

Due to the weather – with increasing cloudiness and chillier temperatures – December isn't an ideal time to catch a meteor shower. But December is the month with the most active meteor shower of the year, called the Geminids. This shower has a ZHR of 150, and the meteors it produces are usually bright and colorful. Unfortunately, a waning gibbous Moon on the peak night will make it difficult to spot faint meteors.
**Parent Object: Asteroid 3200 Phaethon**
**2022 Peak Night: December 13–14**

**Ursids: December 17–26, 2022**

Right on the heels of the Geminids is the often-forgotten Ursid meteor shower. The ZHR for this shower is 10, but sometimes you might catch an outburst that could produce a ZHR of 25. The great news is that the Moon will not interfere during peak night.
**Parent Object: Comet 8P/Tuttle**
**2022 Peak Night: December 22–23**

# The Moon

The Moon is most often the first astronomical object to capture anyone's attention in the night sky. It's also likely the first object most people see up close, whether it be through binoculars or a telescope.

Our nearest celestial neighbor has been an object of fascination since humans first looked up to the sky. Early scientists in ancient observatories carefully tracked the cycles of the Moon. Some 23 centuries ago, Aristarchus of Samos carefully observed a total lunar eclipse and derived impressively accurate measurements of the Moon's diameter. He estimated its diameter was one-third that of Earth. He was close: the actual percentage is 27.2, and its diameter is roughly 3,540 kilometers (2,200 miles).

It takes about 29.5 days for the Moon to go through its cycle of **phases**. When you can't see the Moon at all, it's called a **new Moon**.

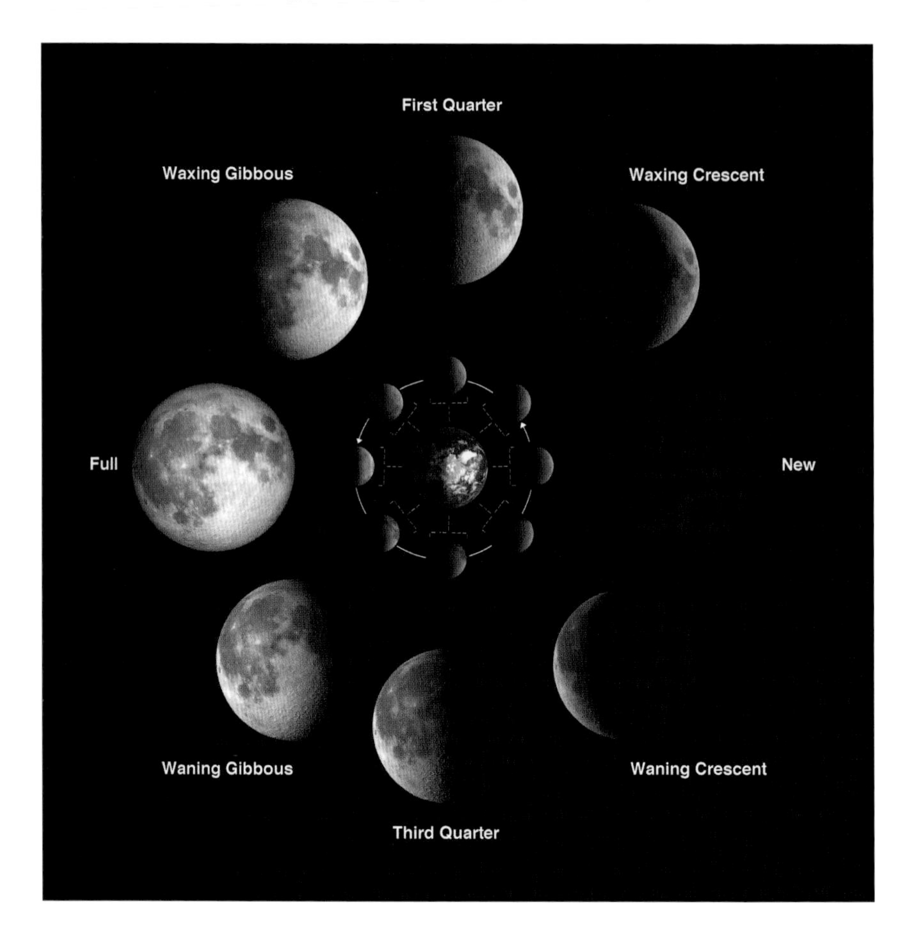

When the Moon is fully illuminated, it is a **full Moon**. The list of phases is new Moon, waxing crescent Moon (when the Moon is just a thin crescent), first quarter Moon (when it appears half full), waxing gibbous Moon (when the Moon is between half-full and full), then full Moon – before the process runs in reverse from full to new: waning gibbous Moon, third or last quarter Moon, waning crescent Moon, new Moon.

The Moon is tidally locked with Earth, meaning that we only ever get to see one face of it. The Sun does shine on the other side of the Moon when we can't see it, so don't call the far side of the Moon the dark side.

While it may seem that we always see the same Moon features month after month, that's not entirely true. The Moon oscillates from our perspective, a process called **lunar libration**. As a result, every so often we see some small part of the Moon that we don't always see.

## Observing the Moon

You don't need a telescope to enjoy the Moon. If observed even briefly on a regular basis, Earth's only natural satellite has much to offer the naked-eye observer. Keep an eye on it, and you will notice things like its wandering path through the constellations, the changing phases, frequent **conjunctions** (where two objects appear close together in the sky) with planets or bright stars, occasional eclipses, lunar libration, earthshine (illumination on the Moon reflected from our planet) and other wonderful atmospheric effects.

If you happen to take a look through a telescope, binoculars or even a camera, our closest astronomical target offers an amazing amount of interesting detail. There are countless features on the lunar nearside, and more than 1,000 have been formally named by the IAU. Many of the greatest names in astronomy, exploration and discovery are commemorated through these names. For extra dramatic effect, try looking at a lunar feature when the **terminator** – the line between darkness and sunlight – runs nearby the feature. The extra shadow could make it easier to see details in the object you're interested in looking at.

An estimated three to four billion years ago, our Solar System was bombarded by debris for hundreds of millions of years. This period is called the Late Heavy Bombardment, and you can see many of the scars from that time left over on the Moon.

Because the Moon has almost no atmosphere and no tectonic activity, those scars have remained over billions of years. Some of that activity created deep **impact basins** that filled with lava flows and hardened into basaltic rock. These largely circular regions were termed **maria**, or seas, by early astronomers because they resembled Earth's oceans.

The brighter white areas, which are highlands that surround the maria and dominate the southern portion of the Earth-facing hemisphere, feature many ancient **impact craters**.

After the bombardment slowed to a trickle, the Moon remained relatively free from intense impact activity – meaning the Moon has changed very little since billions of years ago. But, just like on Earth, meteorites still periodically slam into the Moon today.

The Moon map on the next two pages shows a few interesting features for you to target as you observe the Moon.

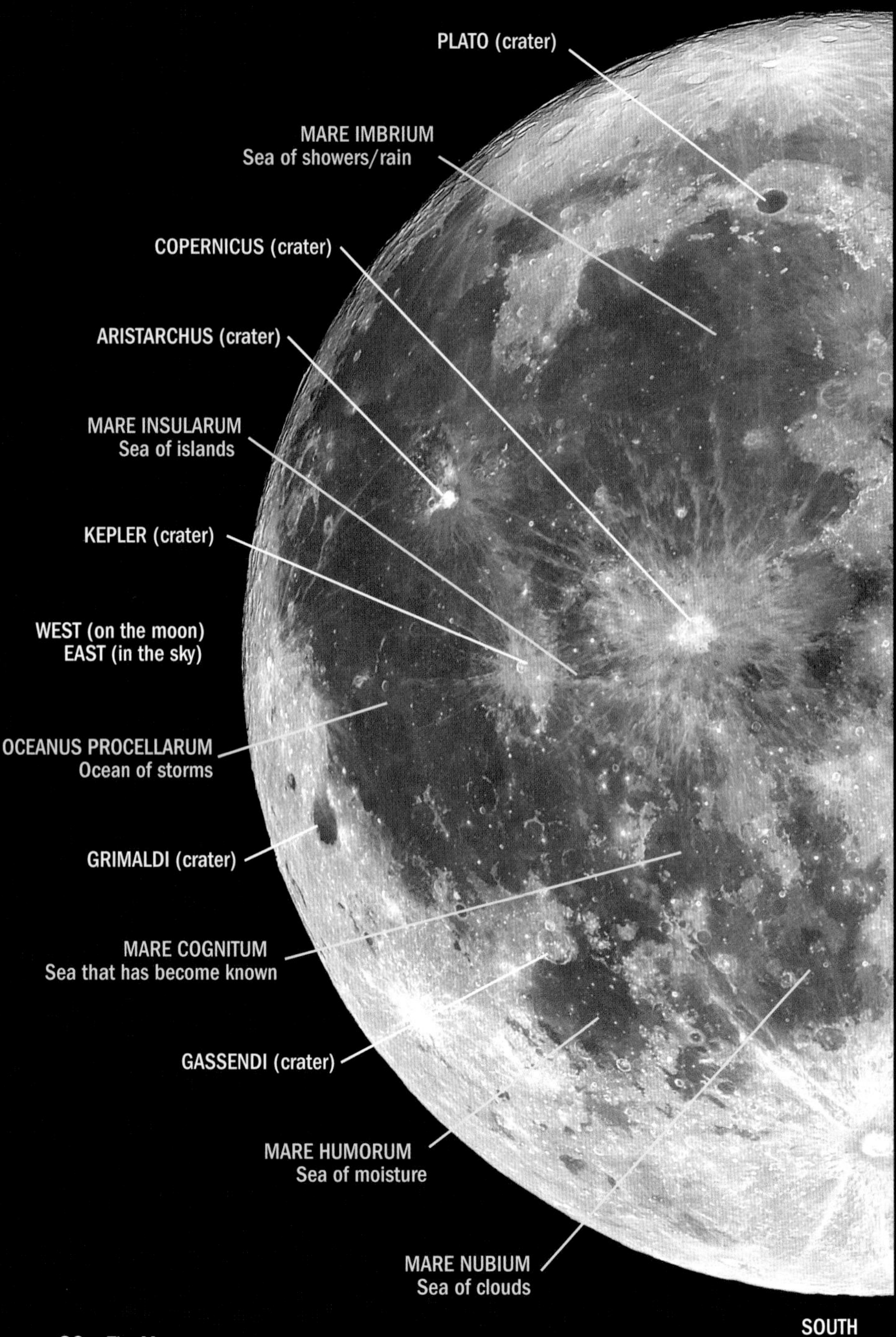
NORTH
PLATO (crater)
MARE IMBRIUM
Sea of showers/rain
COPERNICUS (crater)
ARISTARCHUS (crater)
MARE INSULARUM
Sea of islands
KEPLER (crater)
WEST (on the moon)
EAST (in the sky)
OCEANUS PROCELLARUM
Ocean of storms
GRIMALDI (crater)
MARE COGNITUM
Sea that has become known
GASSENDI (crater)
MARE HUMORUM
Sea of moisture
MARE NUBIUM
Sea of clouds
SOUTH

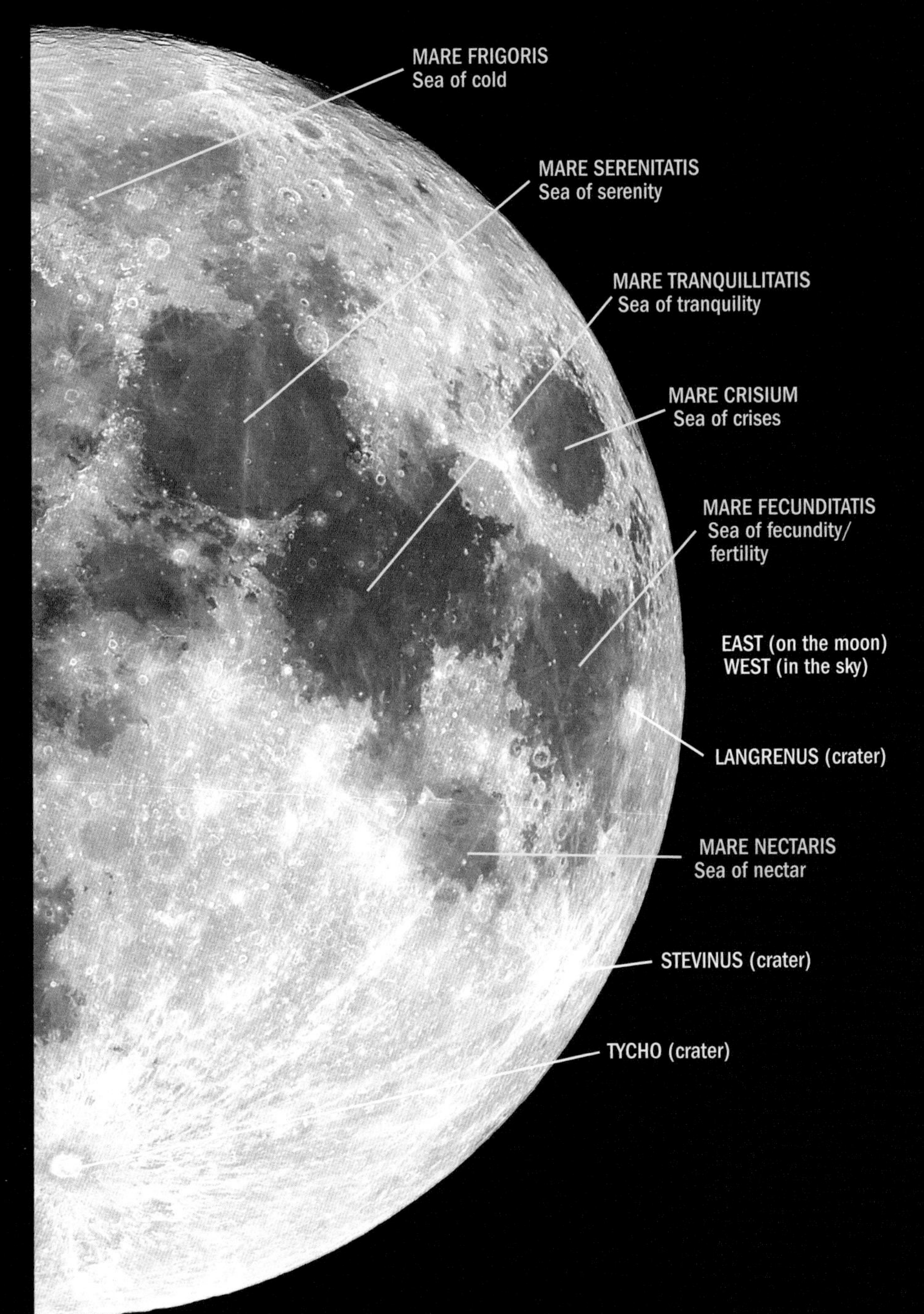
NORTH
MARE FRIGORIS
Sea of cold
MARE SERENITATIS
Sea of serenity
MARE TRANQUILLITATIS
Sea of tranquility
MARE CRISIUM
Sea of crises
MARE FECUNDITATIS
Sea of fecundity/
fertility
EAST (on the moon)
WEST (in the sky)
LANGRENUS (crater)
MARE NECTARIS
Sea of nectar
STEVINUS (crater)
TYCHO (crater)
SOUTH

# The Sun

Our Sun highly influences Earth. It is connected with our seasons, weather, ocean currents and climate, and its power makes life itself possible.

Unlike Earth, the Sun isn't a solid body, and due to that, different parts of it rotate at different rates. At the equator, for example, it spins roughly once every 25 days.

In the core of the Sun, where hydrogen atoms fuse to make helium, temperatures are a searing 15 million degrees Celsius (27 million degrees Fahrenheit). The tremendous amount of energy generated at the core – **thermonuclear fusion** – is carried outward by radiation, taking roughly 170,000 years to get from the core to the top of the convective/

## Take Care of Your Eyesight

Do not observe the Sun without special protective equipment, such as a certified solar filter that covers your eyes or telescope aperture entirely. Unprotected observations even for a few seconds could damage your eyesight permanently.

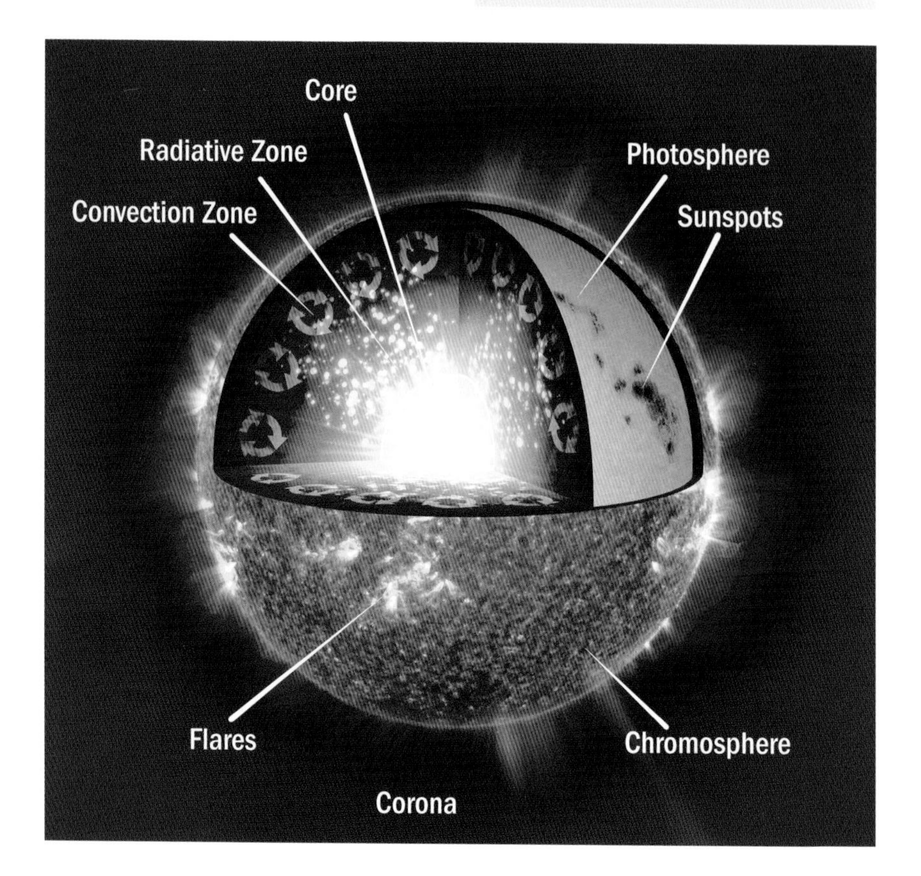

convection zone. As hot as the core is, at the surface, it's a much different story – the temperature is a much "cooler" 5,500 degrees Celsius (10,000 degrees Fahrenheit).

The Sun also has a roughly 11-year cycle, that has a **maximum** (a period of increased solar activity) and a **minimum** (a period of low solar activity). During a maximum, **sunspots**, which are cooler regions on the surface of the Sun, increase. Sometimes, the magnetic field lines of these sunspots can become entangled, finally snapping and releasing a tremendous amount of radiation into space. This process is called a **solar flare**. And often a **coronal mass ejection (CME)** can follow a flare, with charged particles of the Sun speeding outward into space. If these particles reach Earth, they can disrupt radio transmissions, damage satellites and, more positively, interact with our magnetic field. When the particles do interact with the field, they can produce the beautiful Northern Lights or aurora borealis.

Though beautiful, it must be said that these outbursts from the Sun can also cause power outages, as was experienced in Quebec in 1989. As such, astronomers are keen to better understand our nearest star using many spacecraft – such as the Parker Solar Probe, which was launched in 2018. Keep watching the eight-year adventure of the Parker Solar Probe as it studies the Sun's activity.

### A New Solar Cycle

At the end of 2020, we started Solar Cycle 25. After a quiet solar minimum, we can now expect to see more solar activity. Solar Cycle 24 was the weakest cycle in the last 200 years.

The Northern Lights

## Observing the Sun

The Sun, our closest star, is a wonderful object to observe with any type of telescope, if used safely. The only safe filter is the kind that covers the full aperture of the telescope, at the front, allowing only 1/100,000th of the sunlight through the telescope. Without this filter, you risk permanently damaging your telescope or, far worse, your eyes. Ensure that the filter material is specifically certified as suitable for solar observing, and *take great caution every time you observe the Sun.*

The preferred solar filters are made of Thousand Oaks glass or Baader film. Thousand Oaks glass gives the Sun a golden-orange color, while Baader film gives the Sun a more natural white look, which can provide good contrast if there are any bright spots, or **plages**, on the Sun. Other solar-filter options include lightweight Mylar film (which gives the Sun a blue-white tint), and eyepiece projection. Projecting the image from the eyepiece onto white cardstock paper, or a wall, produces an image that can be safely shared with others.

While it might seem like the Sun is just this boring yellow ball in the sky, there are many things to see on its surface, including **prominences**, **filaments**, flares and sunspots. These phenomena result from the strong magnetism within the Sun, which erupts to the surface.

Prominences are solar plasma ejections, some of which fall back to the Sun in the form of teardrops or loops; against the Sun's disk, prominences viewed from above appear as dark filaments. Sunspots are cooler regions of the Sun that appear dark on the face of the surrounding surface. If the highly magnetic lines of sunspots intertwine, they can snap and cause a solar flare – a bright, sudden eruption of energy that can last from a few minutes to several hours. It should be noted that to see prominences and some other solar features, you need very specialized filters.

If you do not have the proper equipment to observe the Sun, you can always visit the Solar and Heliospheric Observatory (SOHO) and the Solar Dynamics Observatory (SDO) websites (see the list of resources on page 120) to see daily images of the Sun.

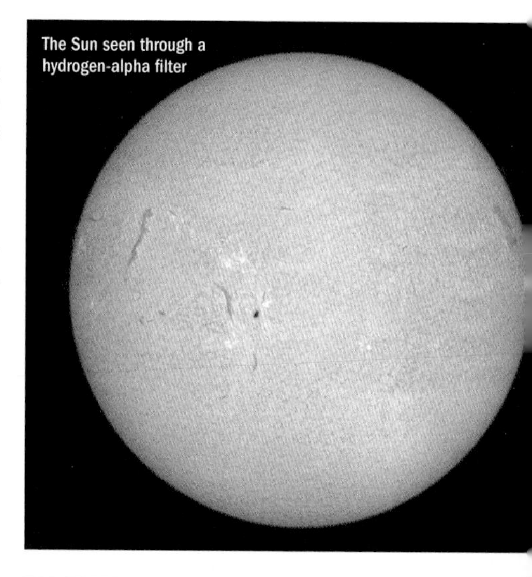
The Sun seen through a hydrogen-alpha filter

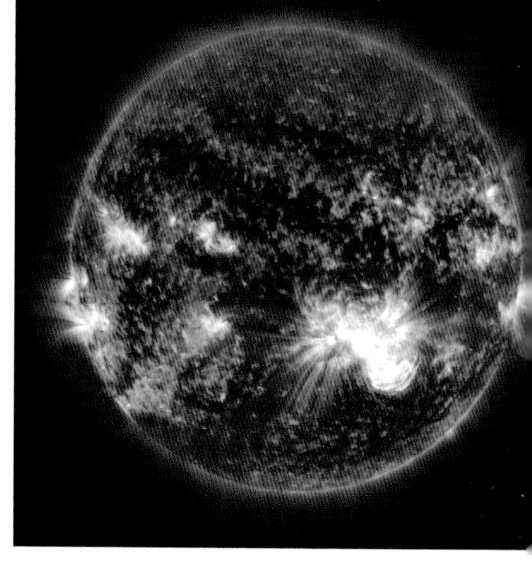
Sun with prominences, filaments, flares and sunspots

# Eclipses

Eclipses – whether they're solar or lunar – are seemingly magical occurrences. In fact, that's exactly what many ancient civilizations believed. Several legends suggested a celestial being devoured the Sun during a solar eclipse. For the Chinese, that being was either a dog or a dragon. The Vikings believed it was two wolves called Hati and Skoll. For the Vietnamese, it was a giant frog. Today, we understand that eclipses are awe-inspiring celestial events resulting from the movement and positions of the Earth, Moon and Sun. There can be as many as seven combined eclipses (i.e. solar and lunar) in a year, but no fewer than four.

A solar eclipse is truly a chance occurrence. The Sun's diameter is roughly 400 times that of the Moon, but the Sun is also about 400 times farther away from Earth than the Moon is from Earth. This means the Moon and the Sun appear roughly the same size in the sky, with only about half a degree across of difference. In a **total solar eclipse**, the new Moon covers the entire face of the Sun, revealing the Sun's stunning corona and prominences. As totality begins and ends, sunlight peeking around the mountainous limb of the Moon creates fleeting effects such as the Diamond Ring and Baily's Beads. If the Moon were farther away or smaller, we wouldn't get this marvelous sight. In fact, the elliptical nature of both the Earth's orbit around the Sun and the Moon's orbit around the Earth means that the apparent diameters of both objects

Solar eclipse

vary significantly. Consequently, it is common for eclipse durations to vary by several minutes.

It should be noted that the totality only occurs in a very narrow path. Outside of this path, observers will only see a partial eclipse. That's why solar eclipse enthusiasts will travel great distances to take in the captivating sight of a totality. Remember, do not look at a solar eclipse unless you use special equipment recommended by The Royal Astronomical Society of Canada or the American Astronomical Society. The only time it is safe to look at a solar eclipse with the naked eye is when it is at its total phase, with the Sun completely covered by the Moon.

Because the Moon orbits at a 5-degree inclination to the ecliptic, we don't get a solar eclipse every month. Every year, there are two 36-day eclipse seasons during which either solar or lunar eclipses can occur, and the eclipse seasons move backward in time every year by 19 days. Also, because the Moon's orbit is elliptical, the angular size of the Moon varies along its path, and sometimes we have an **annular solar eclipse** (or "ring of fire"), when the Moon's disk does not completely cover the Sun's disk. There may be anywhere between zero to two total or annular solar eclipses in a given year. There may also be **partial solar eclipses**, when the Moon's central shadow entirely misses the Earth.

As for lunar eclipses, these occur when Earth is situated directly between the Sun and the full Moon. The Moon drifts through Earth's two shadows: the **penumbra**, which is the fainter, outer shadow, and the deeper shadow that is called the **umbra**. A **penumbral eclipse** is difficult to see with the naked eye as the brightness of the Moon doesn't appear to dim much. But a **partial** or **total lunar eclipse**, when the Moon passes through the umbra, is much more dramatic. During a total lunar eclipse, the Moon can turn an orange-red color, depending on Earth's atmosphere. There may be anywhere between zero to three total lunar eclipses in a given year. Typically, they occur about two weeks before or after a solar eclipse.

While not all eclipses will be visible from North America, you can watch online on sites like SLOOH or The Virtual Telescope Project. You can also see astronomer Fred Espenak's webpage eclipsewise.com for a comprehensive treatment of solar and lunar eclipses.

## Eclipses in 2022

This year, we're only treated to two partial solar eclipses and two total lunar eclipses. Unfortunately, neither of the solar eclipses will be visible from North America. The two lunar eclipses, however, will be visible in at least part of the continent, either partially or in their entirety.

All the times listed are in UTC, or Coordinated Universal Time. See page 40 to learn how to calculate your local time from the time shown in UTC. Also, though we have noted when the Moon enters the penumbra (as this is a part of the stages of a lunar eclipse), it is not noticeable to the human eye.

A total lunar eclipse

## Lunar Eclipses

### Total Lunar Eclipse: May 15–16, 2022

Just two weeks after a partial solar eclipse on April 30 (see page 29), there will be a total lunar eclipse starting on the evening of May 15, and the good news is that it will be visible across a large swath of North America. The entire lunar eclipse, from start to finish, will be seen east of Manitoba, Kansas and Texas. West of those areas, the eclipse will be already underway at Moonrise, but most locations will see the entire duration of totality. The exception is the Pacific northwest, where the Moon will rise in the evening twilight after totality starts.

| Eclipse times (UTC) | |
|---|---|
| Moon enters the penumbra (P1) | 01:33 |
| Partial eclipse begins (U1) | 02:29 |
| Totality begins (U2) | 03:30 |
| Greatest eclipse | 04:12 |
| Totality ends (U3) | 04:55 |
| Partial eclipse ends (U4) | 05:56 |
| Moon exits penumbra | 06:52 |

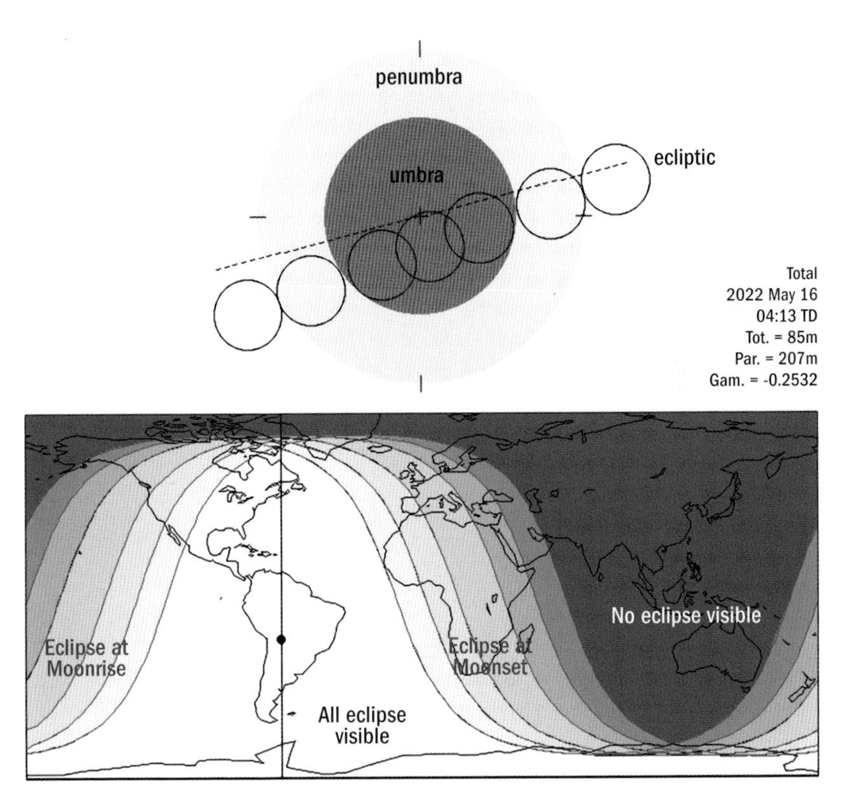


Thousand Year Canon of Lunar Eclipses © 2014 by Fred Espenak

**Total Lunar Eclipse: November 8, 2022**

As is typical with eclipses, two weeks after a solar eclipse (see page 29) we're treated to a lunar eclipse that will be widely visible across North America. This time it's the west that gets to enjoy the entire eclipse from start to finish, particularly Alberta, British Columbia, Yukon, Northwest Territories, Alaska, Washington, Oregon and California. East of those locations, the Moon will set while the eclipse is underway, but most locations will see the entire duration of totality. The exception is for viewers in the Atlantic northeast of the continent, where the Moon will set in morning twilight before totality ends.

| Eclipse times (UTC) | |
|---|---|
| Moon enters the penumbra (P1) | 08:03 |
| Partial eclipse begins (U1) | 09:10 |
| Totality begins (U2) | 10:17 |
| Greatest eclipse | 11:00 |
| Totality ends (U3) | 11:42 |
| Partial eclipse ends (U4) | 12:50 |
| Moon exits penumbra (P4) | 13:57 |

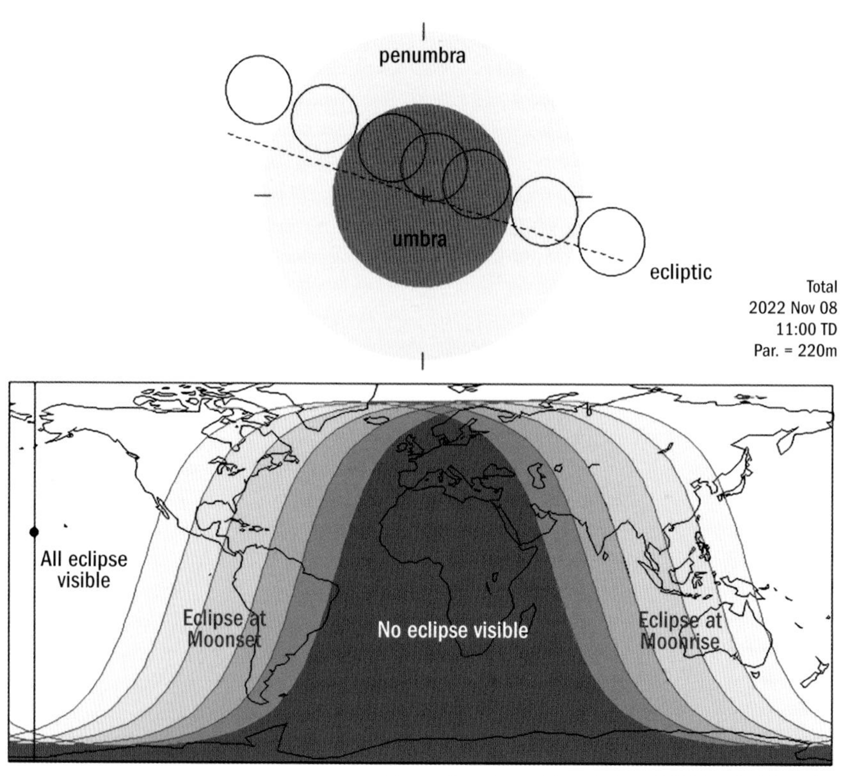


Thousand Year Canon of Lunar Eclipses © 2014 by Fred Espenak

## Solar Eclipses

### Partial Solar Eclipse: April 30, 2022

This partial solar eclipse will be visible from the Southern Sea off the coast of Antarctica and from the Pacific Ocean off the coasts of Chile and Argentina. Some cities in the region might catch it, but it will be low on the horizon. The eclipse begins at 18:46 UTC and lasts until 22:39 UTC, though most of it will occur over open ocean.

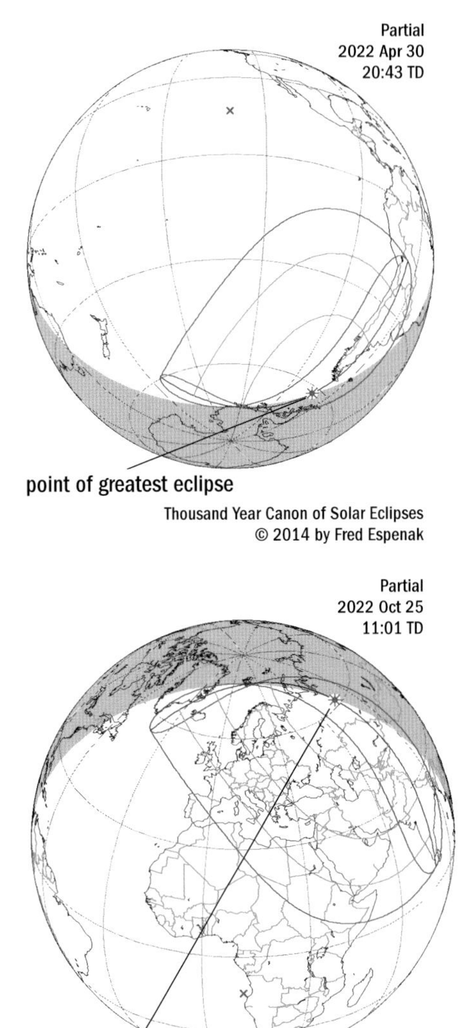


### Partial Solar Eclipse: October 25, 2022

Unlike the eclipse of April 30, this partial solar eclipse will be visible across much of eastern Europe, Asia and the Middle East. It begins at 08:59 UTC and ends at 13:03 UTC. The eclipse maximum will be mainly visible throughout western Russia and Kazakhstan.

# The Northern Lights

The Sun appears to be a bright, unchanging orb in the sky. However, the Sun is anything but unchanging: As we know, it is a ball of constant activity, and that activity has a large influence here on Earth.

One such activity is a solar flare, which is often followed by a coronal mass ejection (CME) that sends particles speeding along the solar wind. Solar flares can reach Earth and disrupt radio transmissions. If Earth is in the path of a CME, the particles can travel down our magnetic field lines toward the poles; this creates the beautiful Northern and Southern Lights, or **aurora borealis** and **aurora australis**, respectively.

As mentioned earlier, the Sun goes through an average 11-year cycle of activity during which it experiences a solar minimum and a solar maximum. In the latter part of 2020, we began a new Solar Cycle, and as the cycle continues, we can expect to see more of the Northern Lights. It's worth noting that over the past few cycles the Sun has been less active than in the past.

Auroras come in different shapes and sizes. They can be steady, moving or rapidly pulsating. They also come in an array of colors, depending on how the particles interact with molecules at different altitudes. They are also hard to predict and can be difficult to see. Catching them is a special treat, even for experienced astronomers. Be aware that to the unaided eye the auroral colors are usually muted, as the light is not strong enough to stimulate our color vision. The bright colors seen in photographic reproductions of auroras show up when long exposures are used.

Green is the most common color and occurs when particles interact with oxygen molecules at an altitude of roughly 100 to 300 kilometers (62 to 186 miles). Between 300 to 400 kilometers (186 to 248 miles) the oxygen molecules produce a red color, and below 100 kilometers (62 miles) the molecules interact with nitrogen – producing a pink color.

While the interaction of these solar particles can produce a beautiful light show, they can also be destructive.

One of the most powerful events from a CME was called the Carrington Event. In 1859, English astronomers Richard Carrington and Richard Hodgson were the first to ever witness a solar flare. But when the particles reached Earth, they disrupted telegraph systems in North America and Europe, in some reports even setting equipment on fire. The Northern Lights were visible as far south as Honolulu and the Southern Lights as far north as Santiago, Chile.

A massive CME, like the one that caused the Carrington Event, has a real chance of disrupting GPS and satellite communications and destroying electrical grids. As a result, space agencies have been launching more satellites and probes to the Sun to monitor space weather, and power companies have been developing contingency plans to ensure the next powerful solar eruption doesn't cause such damage.

The Northern Lights over Saskatchewan

boasts the most spectacular storm formation in our Solar System, called the Great Red Spot. Jupiter orbits 772 million kilometers (479 million miles) from the Sun and a single day is about 10 Earth hours long. It orbits the Sun in about 12 Earth years.

Jupiter is the second-brightest planet in our night sky and is also one of the most enjoyable planets to observe. With a modest telescope, you can easily see the cloud bands in its atmosphere. But you can even enjoy the planet through a pair of binoculars; if you watch the planet every night, four of its 79 confirmed moons – Ganymede, Io, Europa and Callisto – can be seen changing positions night after night.

The Great Red Spot (GRS) is a massive, swirling egg-shaped storm that has been shrinking for reasons that are unclear to astronomers. In the 1800s, the GRS was estimated at 41,000 kilometers (25,476 miles) along its long axis. NASA's Juno spacecraft measured the GRS at 16,350 kilometers (10,159 miles) in width on April 3, 2017. Several astronomy apps predict the best time on any night for viewing the GRS through a telescope.

Jupiter

Jupiter spends most of 2022 in Pisces and is in opposition on September 26 at magnitude -2.9. And if you're willing to get up before sunrise, there will be a beautiful conjunction of Jupiter and Venus in the east. The pair will be roughly one-third of a degree apart, with Mars lying nearby to the right.

With its magnificent rings, **Saturn** is often considered the jewel of our Solar System. With a pair of binoculars, Saturn's pancake shape is evident. Look through a telescope, even a modest one, and its fine rings are clearly visible.

Saturn is nine times wider than Earth and 1.4 billion kilometers (886 million miles) away from the Sun. This gaseous planet has the second-shortest day in the Solar System, lasting roughly only 10.7 hours, while its orbit takes 29.4 Earth years. The planet also has more than 80 moons, though its largest, Titan, is the only one easily viewed through a modest telescope.

Saturn is the second-largest planet in the Solar System. It's not the only planet to have rings (Jupiter, Uranus and Neptune all have ring systems), but it has by far the most spectacular ring system we can see. Saturn's rings are made up of billions of pieces of ice that range from tiny dust-sized grains to chunks as big as a house. The ring system is roughly 282,000 kilometers (175,000 miles) across but only about 9 meters (30 feet) tall. There are several rings, mostly close to each other, but the main rings are called A, B, and C. The gap between the outer ring (A) and the first inner ring (B) is called the Cassini Division, following the discovery by Italian astronomer Giovanni Cassini.

Saturn spends all of 2022 in Capricornus and is at opposition on August 14 at magnitude +0.3. On April 5, Saturn and Mars have a beautiful conjunction just before sunrise, when they will be less than one-third of a degree apart.

Saturn

**Uranus** is the faintest of the planets visible to the naked eye (in dark-sky conditions), and the third-largest planet overall. This giant ice planet has 13 faint rings and 27 small moons. But most interestingly, Uranus rotates on its side, at nearly a 90-degree angle.

In 1781, Uranus was the first planet to be discovered with the aid of a telescope by astronomer William Herschel. However, Herschel first believed it was either a star or a comet. It was confirmed as a planet two years later by Johann Elert Bode.

The planet is four times the size of Earth. A day on Uranus takes roughly 17 Earth hours, with one orbit taking 84 Earth years. Uranus orbits on average 2.9 billion kilometers (1.8 billion miles) from the Sun.

Just as it did in 2021, Uranus spends all of 2022 in Aries, reaching opposition on November 9 at magnitude +5.6.

At an average distance of 4.5 billion kilometers (2.8 billion miles) from the Sun, **Neptune** is the farthest planet in our Solar System, taking about 165 Earth years to complete one orbit. Like Uranus, Neptune is roughly four times larger than Earth, and it, too, has a faint ring system.

Neptune also holds the distinction as being the windiest planet in our Solar System, with clouds of frozen methane being whipped across the planet at 2,000 kilometers (1,200 miles) per hour. A day on Neptune is about 16 Earth hours long.

Neptune spends most of 2022 in Aquarius, never wandering too far from Jupiter. It reaches opposition on September 16 at magnitude +7.8.

# Deep Sky Objects

The Universe has many amazing sights to behold, including stunning clusters of stars, nebulae and swirling galaxies.

There are two main types of star clusters: **open** and **globular**. Globular clusters are old star systems at the edge of spiral galaxies that can contain anywhere from thousands to millions of stars, packed in a close, roughly spherical form and held together by gravity.

Two beautiful globular clusters you can see from the Northern Hemisphere include Messier 13, found in the constellation Hercules (a hero from Greek mythology), and Messier 3 in the constellation Canes Venatici (the hunting dogs).

Open clusters contain anything from a dozen to hundreds of stars, but the stars are more spread out and found on the **galactic plane**, the plane on which most of a galaxy's mass lies. Perhaps the most famous open star cluster, the Pleiades (Messier 45) can be spotted with the naked eye and through light pollution.

As well, there are **nebulae** – clouds of dust and gas left over from a **supernova**, an exploding star. These are considered "stellar nurseries," as eventually the gas and dust will coalesce into new stars and potential stellar systems with planets, moons and, possibly, life. One of the most famous and easily visible is the Orion Nebula (Messier 42), found in the winter constellation Orion.

There are also **emission nebulae**, which are clouds of interstellar gas excited by nearby stars that emit their own light at optical wavelengths. **Planetary nebulae** are cloudy remnants left over from stars that shed their gas and dust late in their lives. And finally, there are **dark nebulae**, interstellar clouds that are so dense they obscure the light of the objects behind them.

Messier 13, a globular cluster

## Messier Catalog

The Messier Catalog is a register of 110 objects in the night sky, including open and globular clusters, nebulae, galaxies and one double star. The catalog was started by Charles Messier in the 18th century. Messier was chiefly interested in finding comets, and the catalog is his list of non-comet objects that he observed. You can find a list of all the objects in the Messier Catalog on pages 114–117.

The Pleiades (Messier 45)

The Orion Nebula (Messier 42)

## Galaxies

Galaxies come in many different shapes and sizes. There are four main types, however: elliptical, spiral, barred spiral and irregular. The diagram below shows Edwin Hubble's scheme for classifying galaxies, colloquially known as the "tuning fork" because of its shape.

**Elliptical galaxies** seem somewhat disorganized and look roughly egg-shaped. Shapes range from almost circular (E0) to very elliptical (E7).

**Spiral galaxies** have both a large central bulge and a thin disk of stars. Much like elliptical galaxies, these also range from tight spirals (Sa) to more diffuse (Sd).

**Barred spiral galaxies** are similar to spirals, but they have visible "arms" or bars near the center, and they range from tight (SBa) to diffuse (SBd).

There are also disk galaxies that don't have spiral arms. These are called **lenticular (lens-shaped) galaxies**. They are classified as S0.

The most common galaxy is the spiral, accounting for more than 75 percent of galaxies in the visible Universe. Our galaxy, the Milky Way, is believed to be a barred spiral. Our closest neighboring spiral galaxy is the Andromeda Galaxy (Messier 31), which is easily visible through binoculars or with the naked eye in dark skies.

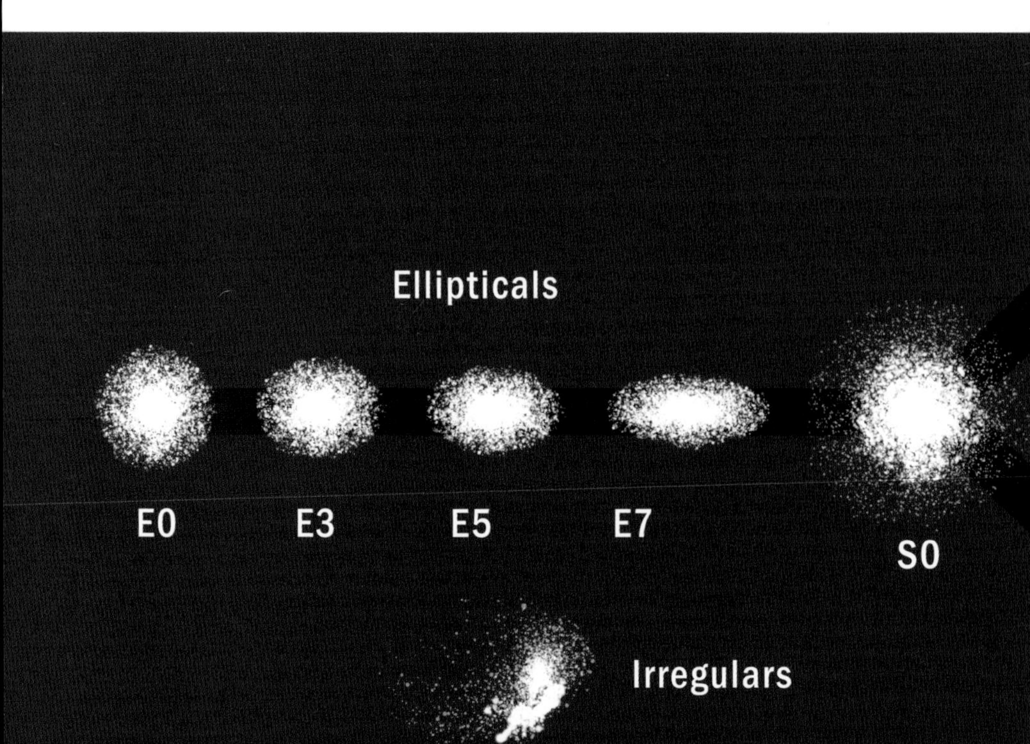

The Andromeda Galaxy (Messier 31)

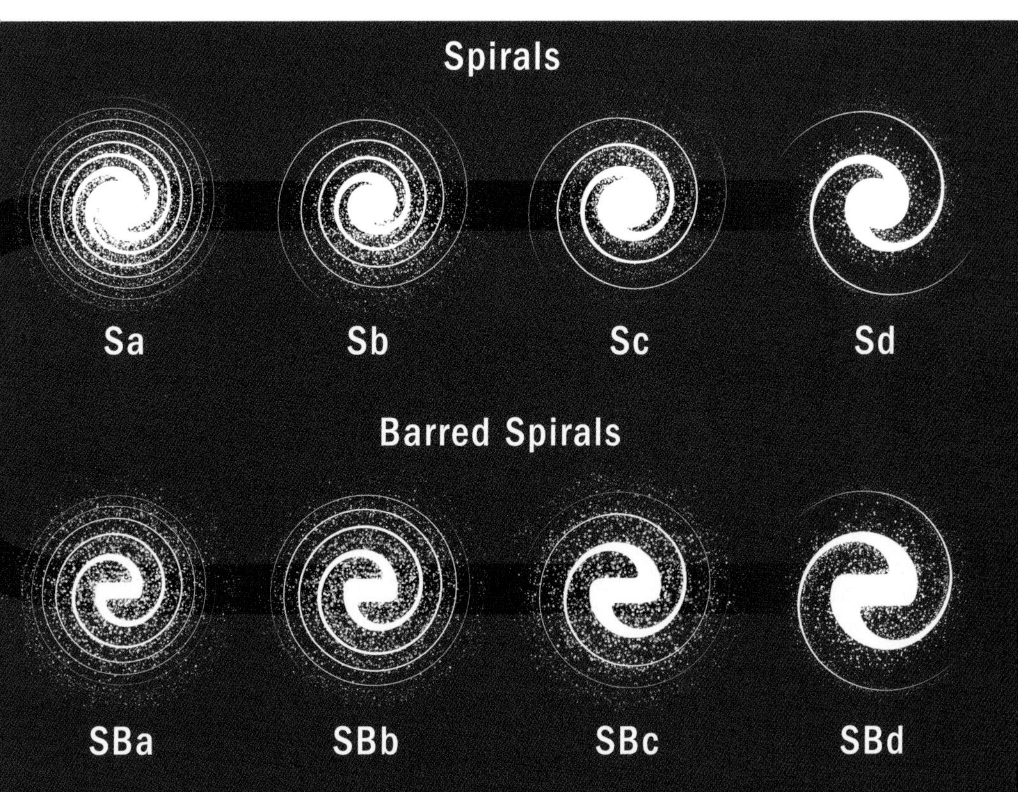

# The Sky Month-by-Month

## Introduction

The following pages are your guide to the sky for every month in 2022. Each month features a calendar of events, information about Moon phases, sky charts facing south and north and descriptions of interesting objects to target.

### Monthly Events

This section summarizes the events of each month, including Moon phases, conjunctions between the Moon and planets, conjunctions between planets and other planets, oppositions, eastern and western elongations, meteor shower peaks and much more.

All times are listed in UTC, or Coordinated Universal Time, which is the standard time used by astronomers throughout the year. The map shown below will guide you on how to calculate your local time from the time shown in UTC. It's important to note that most of North America will observe Daylight Saving Time (DST) between March 13, 2022, and November 6, 2022. Between these two dates, you will need to add an additional hour to your local time.

You'll notice some of the times listed fall during daylight, when observation is likely impossible. However, you can use this date

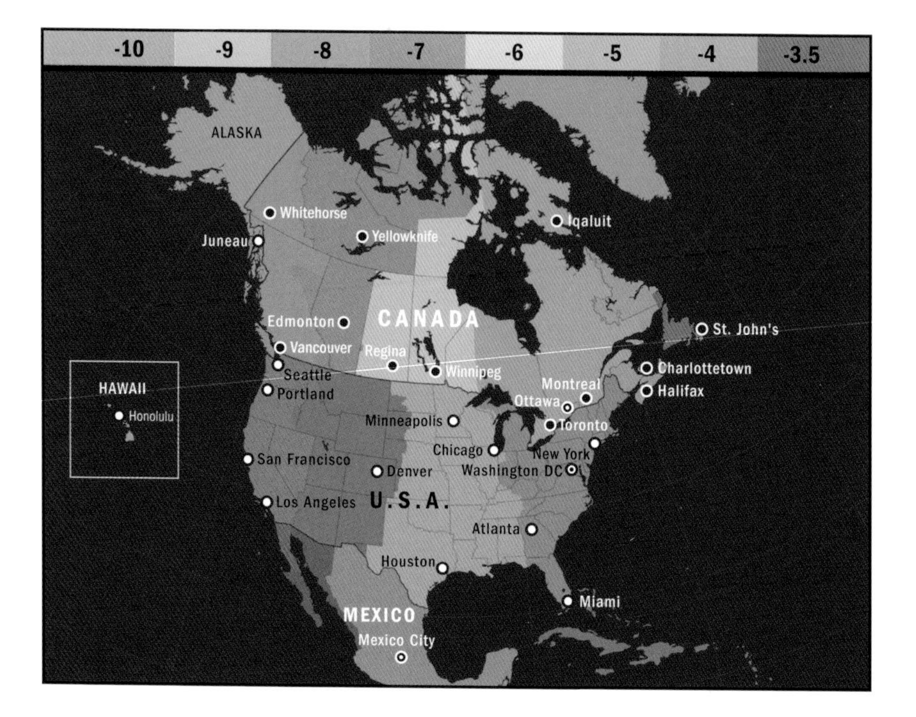

and time as a guide for observing an event on either the preceding night or the following night. Similarly, the coordinates given as the distance between celestial objects should be used as a guide and will not reflect exactly what you will see in the night sky. This is particularly true for close approaches that take place during the day, local time. For example, the calendar of events might say Mercury is 1.5 degrees south of Jupiter at 21:00 UTC, but by the time Mercury and Jupiter become visible to an observer in Eastern Time, that separation might have increased to 2.4 degrees.

This section also features a calendar layout with the Moon phases for each day shown, as well as a description of interesting Moon events for the month. You might notice some discrepancies between the coordinates given in these descriptions and those listed in the table of events. Because the Moon moves in right ascension so quickly, conjunctions between the Moon and planets as listed in the table may appear different for North American observers, and the separation between the objects may be more than suggested. The coordinates in the descriptions have been altered to more closely reflect the view from North America, but of course there may be more variation based on when the objects are visible to you in the night sky.

The Moon phases shown in the calendar are based on the day they occur in UTC time. For some observers in North America, that might mean the ideal time to watch, for example, the full Moon rising would be the night before.

## Sky Charts and Descriptions

Each month features two sky charts, one facing south and the other facing north. These charts highlight select constellations, stars, clusters, nebulae, planets and galaxies visible in the night sky. While these sky charts are an accurate representation of the sky, it should be noted that factors like light pollution, smoke, cloud cover and so on can affect how much you see, and some of the fainter stars shown on the charts may not be visible.

The charts are drawn at a latitude of 45 degrees north. Observers north or south of this latitude will see slightly more of the northern or southern sky, respectively. The charts show the sky at 10:00 p.m. local standard time on the 15th of each month. You will also have the same view of the sky at 11:00 p.m. local standard time at the beginning of each month and at 9:00 p.m. local standard time at the end of each month. We note these times on the sky charts and also include Daylight Saving Time in parentheses between March and November. The planets shown on the sky charts will move slightly relative to the stars over the course of the month.

The preceding month's charts can be used for sky viewing two hours earlier, and the following month's charts can be used for sky viewing two hours later. So, for example, if you wanted to view the sky at 8:00 p.m. local time in February, you would refer to January's sky chart. If you're referring to a previous or later month's sky chart, note that the positions of the planets will not be correct.

# January

## January's Events

January can be the coldest month for many parts of North America, but if the skies are clear, it offers some of the best views of the starry sky. The nights are long, the air is free from summertime humidity and the sky is just waiting for you to bundle up and grab a pair of binoculars (or not!), so you can take in the marvels of the winter sky. It's also time for one of the best meteor showers of the year, the Quadrantids.

The constellation Orion

## Calendar of Events

| Day | Time (UTC) | Event |
|---|---|---|
| 1 | 23:00 | Moon at perigee: 358,000 km (222,451 mi.) |
| 2 | 18:33 | New Moon |
| 3–4 | | Quadrantid meteor shower peak |
| 4 | 06:55 | Earth at perihelion 0.9833 astronomical units (AU) |
| 4 | 16:50 | Saturn 4°N of Moon |
| 6 | 00:09 | Jupiter 4°N of Moon |
| 7 | 11:04 | Mercury 19°E of Sun (greatest eastern elongation) |
| 9 | 18:11 | First quarter of the Moon |
| 13 | 04:17 | Mercury 3.4°N of Saturn |
| 14 | 09:27 | Moon at apogee: 405,800 km (252,152 mi.) |
| 17 | 23:48 | Full Moon |
| 25 | 13:41 | Last quarter of the Moon |
| 29 | 15:05 | Mars 2°N of Moon |
| 30 | 07:09 | Moon at perigee: 362,300 km (225,123 mi.) |

## The Moon This Month

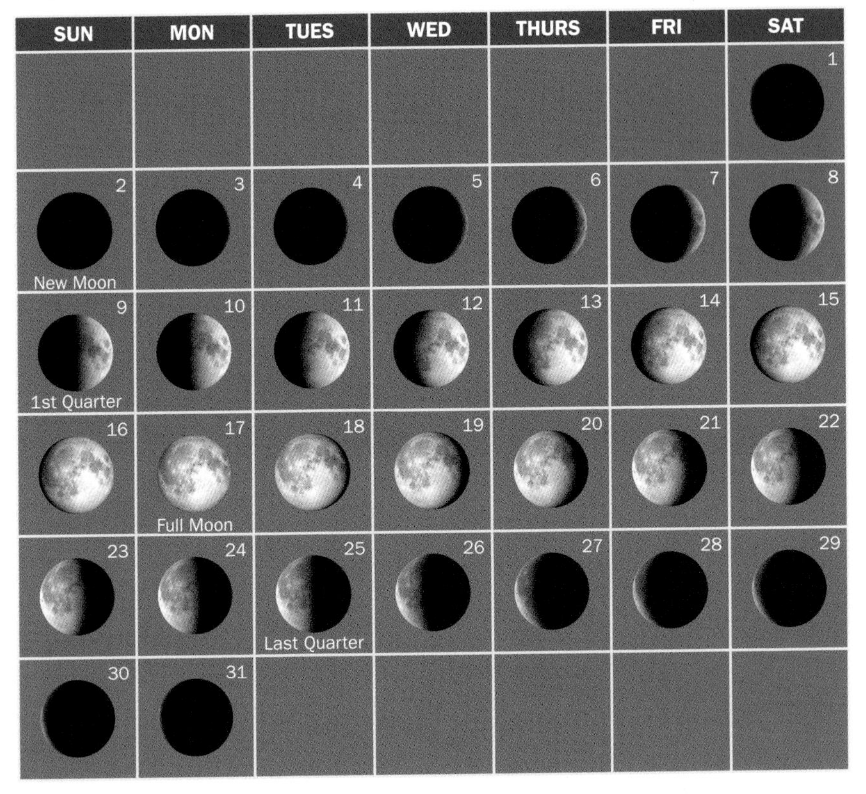


The month – and the new year – starts off with the waning crescent Moon barely illuminated at roughly 2 percent, with the new Moon falling on January 2. This is particularly good news, as the Quadrantid meteor shower peaks on the night of January 3 to 4. No moonlight means that faint meteors will be easier to see, even in light-polluted cities (though it's always best to head out to the darkest location you can).

On January 4, the crescent Moon appears near Saturn very low in the southwest in the evening twilight. On the evening of January 5, the Moon sits south of Jupiter. Look for Mercury in the evening twilight during the first half of the month. It has its greatest elongation east of the Sun on January 7 and is within a binocular field of Saturn for about a week after that. Venus becomes visible in the morning twilight beginning on January 15, low on the eastern horizon.

On January 17, the full Moon will sit between the constellations Cancer and Gemini. You can find it just below the twin "head" of Pollux.

On January 29, if you want to challenge yourself, get up early and look low toward the southern horizon, where you can find the crescent Moon just below Mars with Venus shining brightly nearby in the morning twilight.

The Moon hits perigee both on January 1 and January 30. It will be at apogee on January 14.

URSA MAJOR
Dubhe
Merak
CEPHEUS
CAMELOPARDALIS
LACERTA
Caph
CASSIOPEIA
Double Cluster
LYNX
LEO MINOR
α Lyn
Mirfak
PERSEUS
Capella
Zenith
AURIGA
Andromeda Galaxy
ANDROMEDA
Almach
Mirach
Alpheratz
PEGASUS
Castor
Pollux
GEMINI
Flaming Star Nebula
TRIANGULUM
Triangulum Galaxy
Elnath
Pleiades
Merope Nebula
ARIES
Hamal
LEO
Algieba
CANCER
Beehive Cluster
Alhena
Aldebaran
Hyades
TAURUS
Uranus
PISCES
Regulus
CANIS MINOR
Betelgeuse
Bellatrix
Ecliptic
Procyon
Menkar
ORION
E
SEXTANS
HYDRA
MONOCEROS
Rigel
CETUS
W
Orion Nebula
ERIDANUS
Sirius
Alphard
CANIS MAJOR
LEPUS
Diphda
PUPPIS
Adhara
FORNAX
CONSTELLATIONS
CÆLUM
Galaxies
COLUMBA
SE
SW
Star Clusters
Nebulæ
S
Planets
Stars
Region of interest
January 1: 11:00 p.m.
January 15: 10:00 p.m.
February 1: 9:00 p.m.

## Highlights in the Southern Sky

We start off with none other than the most prominent winter constellation: **Orion (the Hunter)**. When Orion is visible, the seven brightest stars in the night sky are as well: **Betelgeuse**, **Rigel**, **Aldebaran**, **Capella**, **Pollux**, **Procyon** and **Sirius**. The mythical hunter hangs high in the southern sky with what is probably one of the most prominent stars, Betelgeuse, at Orion's left shoulder. Betelgeuse is most remarkable for its reddish color. It is a red supergiant that is nearing the end of its life – in about 100,000 years or so – and it is massive: It's about 1,400 times larger and 4,000 times more luminous than our Sun.

Another prominent star in the constellation is Rigel, the bright right foot of Orion. You might also notice a particularly bright star east of the hunter's left foot. That would be Sirius, the brightest star in the night sky and part of the constellation **Canis Major (the Great Dog)**.

Astronomers believe that many stars are part of multiple star systems. Many of those are binary stars, which are a pair of stars that orbit a common center of mass. Sirius just happens to be one of those stars. There's Sirius A, which is the bright one we see, and Sirius B, which is a small blue-white star that lies roughly 8.6 light-years from Earth. Sirius B is hotter than Sirius A and smaller than the planet Earth – it was the first white dwarf star to be discovered.

But Orion isn't the only favorite of the southern sky. There's also **Gemini (the Twins)**, which lies just to the left of Orion in the evening sky. The constellation's most prominent stars are **Castor** and Pollux, the heads of the mythological Greek twins.

**Auriga (the Charioteer)** is also prominent high in the southwest, just above Orion. The most notable star is Capella, which shines brightly as the sixth-brightest star in the night sky. By mid-month, it lies almost directly overhead, at the zenith. If you happen to have a small telescope and want a challenging object, you can look in the middle of the constellation and see the amazing **Flaming Star Nebula (IC 405)**. It is about 1,500 light-years from Earth and 5 light-years across. This nebula makes for an amazing photographic target.

Flaming Star Nebula (IC 405)

Using binoculars to scan the winter Milky Way from Auriga to **Perseus** to **Cassiopeia**, you will discover several open clusters of stars, including the famous **Double Cluster (NGC 884 and NGC 869)**.

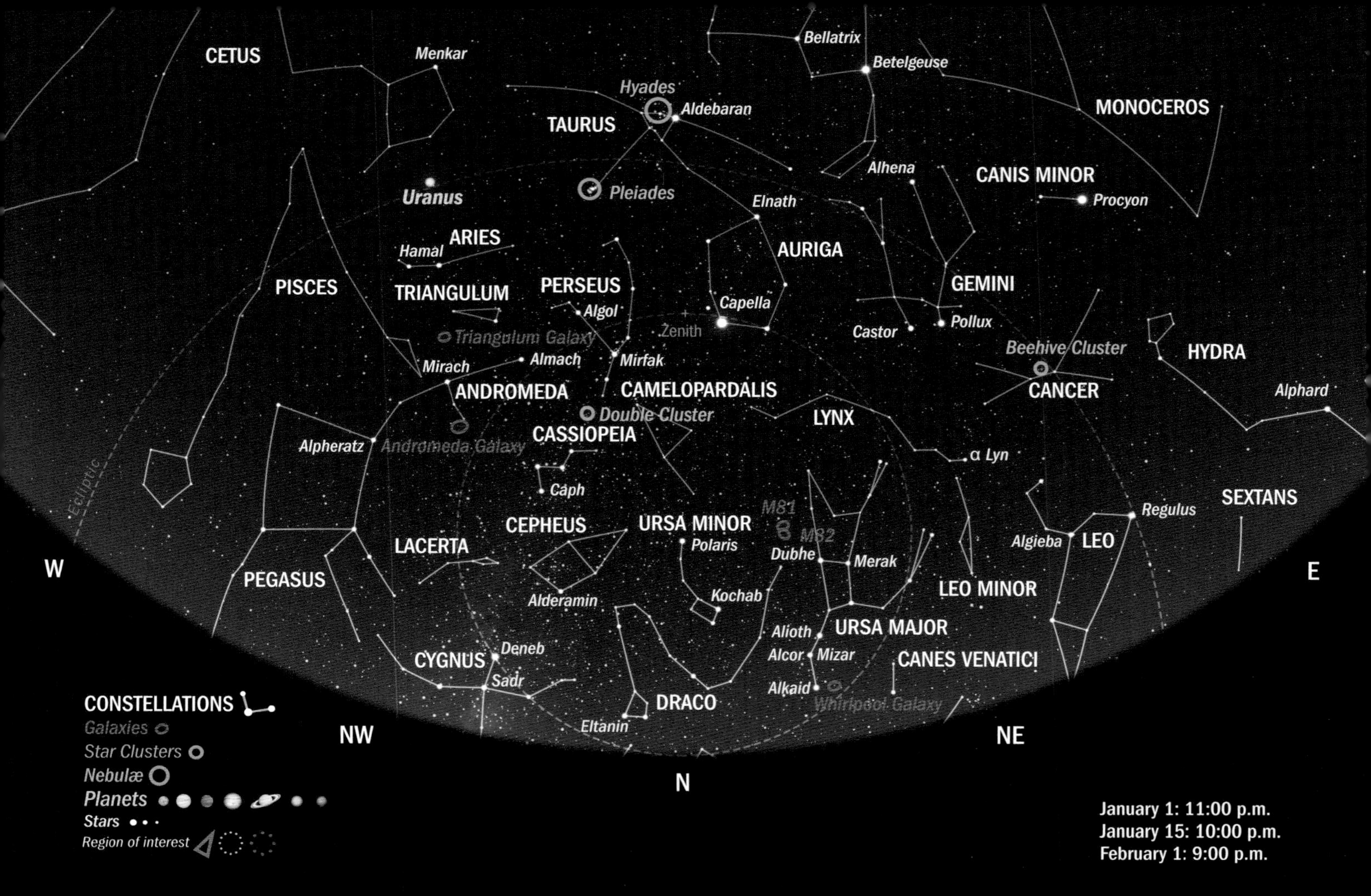
CETUS
Menkar
Hyades
Aldebaran
TAURUS
Bellatrix
Betelgeuse
MONOCEROS
Uranus
Pleiades
Elnath
Alhena
CANIS MINOR
Procyon
ARIES
Hamal
PISCES
TRIANGULUM
PERSEUS
Algol
AURIGA
Capella
GEMINI
Castor
Pollux
Zenith
Triangulum Galaxy
Mirach
Almach
Mirfak
Beehive Cluster
HYDRA
ANDROMEDA
CAMELOPARDALIS
CANCER
Alphard
Double Cluster
LYNX
Alpheratz
Andromeda Galaxy
CASSIOPEIA
α Lyn
Ecliptic
Caph
M81
M82
SEXTANS
Regulus
CEPHEUS
URSA MINOR
Polaris
LACERTA
Dubhe
Merak
Algieba
LEO
W
PEGASUS
E
Alderamin
Kochab
LEO MINOR
URSA MAJOR
Alioth
Deneb
Alcor
Mizar
CANES VENATICI
CYGNUS
Sadr
Alkaid
DRACO
Whirlpool Galaxy
CONSTELLATIONS
Galaxies
Star Clusters
Nebulæ
Planets
Stars
Region of interest
Eltanin
NW
NE
N
January 1: 11:00 p.m.
January 15: 10:00 p.m.
February 1: 9:00 p.m.

## Highlights in the Northern Sky

If we started off with Orion in the southern sky, then we should probably start off with none other than **Ursa Major (the Great Bear)** in the northern sky. Ursa Major lies in the northwest this month. The constellation is actually best known for its asterism, the **Big Dipper**. Though it looks like a group of seven stars make up this prominent asterism, it's actually eight. If you look very carefully at the second star in the handle, you'll notice that it's actually two stars, **Alcor** and **Mizar**. According to legend, ancient Romans and the Arabs would use the pair as a visual acuity test. The Mi'kmaq First Nation see Mizar as the Chickadee and Alcor as his cooking pot from the story of the great bear and the seven hunters. In January evenings, you can find the Big Dipper standing upright on its handle, with the bowl pointing west.

The Big Dipper is a handy way to find the celestial due north, which is marked by **Polaris**, otherwise known as the "North Star." Polaris is part of the constellation **Ursa Minor (the Little Bear)** and marks the end of the handle of the **Little Dipper** asterism, a smaller version of the Big Dipper. Using the Big Dipper as a guide, find the two stars at the top of the bowl, **Merak** to the right and **Dubhe** to the left. Trace a line from Merak to Dubhe and beyond, and you'll find Polaris. In the Northern Hemisphere, the North Star appears almost stationary in the sky night after night, month after month. If you were to set up a camera pointing directly at Polaris, you'd notice the constellations moving around it as the night progresses.

Two of the most famous galaxies are **Messier 81 (Bode's Galaxy)** and **Messier 82 (the Cigar Galaxy)**. M81 can be seen from dark-sky sites through binoculars or a small telescope, and M82 is in the same field of view.

The **Quadrantid meteor shower**, which actually began on December 27, 2021, peaks on the night of January 3 to 4. The Quadrantids are one of the year's most active showers with a ZHR of 120. The radiant, or the direction from which the meteors appear to originate, lies just beneath the star **Alkaid**, at the tail of Ursa Major, but of course you only need to look up to catch a few "shooting stars."

Messier 81 and Messier 82

# February

## February's Events

The Great Orion Nebula (Messier 42)

The days are starting to get noticeably longer now, but on the flip side, the cloud-covered nights are also diminishing, particularly for the more northern parts of the continent. The mighty Orion is still prominent in the southern sky, though by the end of the month it is placed firmly in the southwest.

This month is a great time to pull out those binoculars, if you have them, to check out some of the beautiful star clusters and nebulae that grace the night sky. One of the best binocular targets is undoubtedly Messier 42, or the Great Orion Nebula. You won't be disappointed.

If you'd like to check out a few planets, get up before sunrise to catch a trio: Mercury, Venus and Mars are all visible low in the east before sunrise, though by the end of the month, Mercury will all but disappear as it becomes washed out by the rising Sun.

## Calendar of Events

| Day | Time (UTC) | Event |
|---|---|---|
| 1 | 05:46 | New Moon |
| 2 | 21:08 | Jupiter 4°N of Moon |
| 7 | 19:39 | Uranus 1.2°N of Moon |
| 8 | 13:50 | First quarter of the Moon |
| 11 | 02:39 | Moon at apogee: 404,900 km (251,593 mi.) |
| 16 | 21:07 | Mercury 26°W of Sun (greatest western elongation) |
| 16 | 16:57 | Full Moon |
| 23 | 22:32 | Last quarter of the Moon |
| 26 | 22:18 | Moon at perigee: 367,800 km (228,540 mi.) |
| 27 | 06:00 | Venus 9°N of Moon |
| 27 | 09:00 | Mars 4°N of Moon |

## The Moon This Month

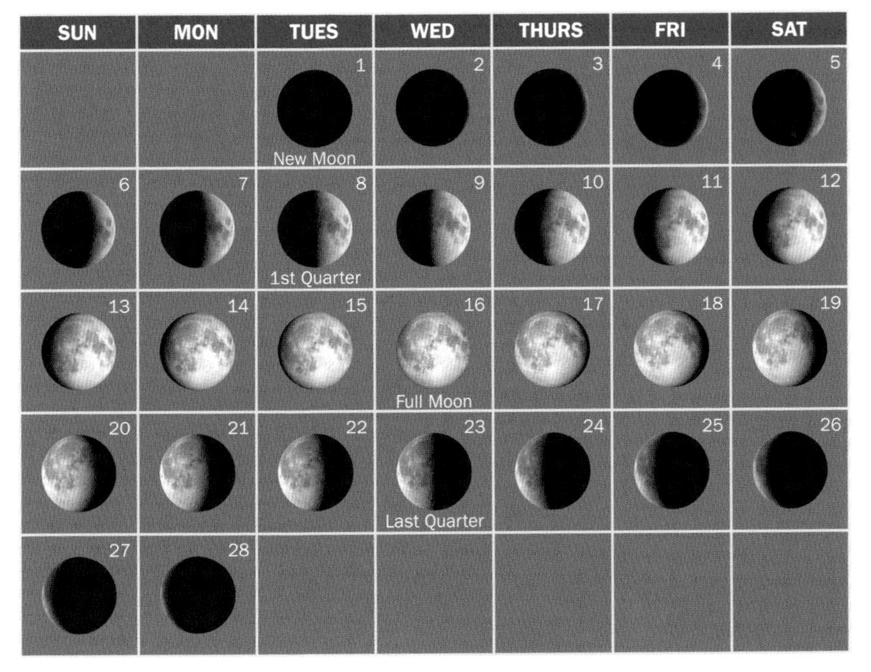

| SUN | MON | TUES | WED | THURS | FRI | SAT |
|---|---|---|---|---|---|---|
| | | 1 New Moon | 2 | 3 | 4 | 5 |
| 6 | 7 | 8 1st Quarter | 9 | 10 | 11 | 12 |
| 13 | 14 | 15 | 16 Full Moon | 17 | 18 | 19 |
| 20 | 21 | 22 | 23 Last Quarter | 24 | 25 | 26 |
| 27 | 28 | | | | | |

While January started off with an *almost* new Moon, February begins with a proper new Moon, illuminated by about 0.5 percent. On January 2, a slim crescent Moon joins Jupiter after sunset, low in the southwestern sky. If you have never seen the planet Uranus, look for it in binoculars on the night of February 7 to the right of the first quarter Moon.

The Moon reaches apogee on February 11, with the full Moon falling on February 16. Perigee occurs on February 26. Mercury reaches its greatest elongation west on January 26, and it sits very low in the southeast before sunrise for several days.

For something of a treat, on February 27 Venus and Mars may be visible, but the very thin waning crescent Moon will be all but invisible, rising in daylight.

A stunning halo surrounds the Moon

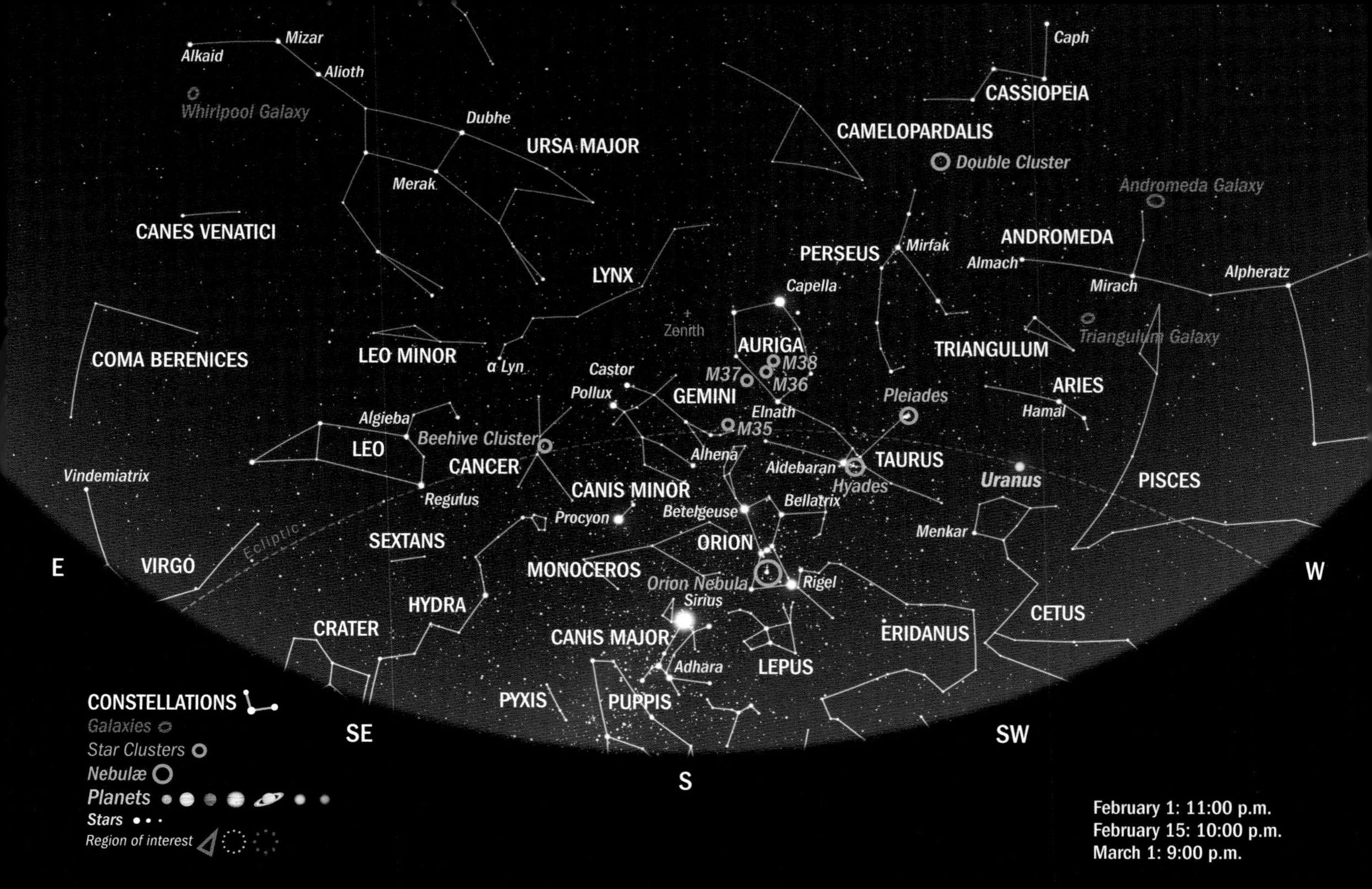

Mizar
Alkaid
Alioth
Whirlpool Galaxy
Dubhe
URSA MAJOR
Merak
CANES VENATICI
LYNX
CASSIOPEIA
Caph
CAMELOPARDALIS
Double Cluster
Andromeda Galaxy
ANDROMEDA
PERSEUS
Mirfak
Almach
Mirach
Alpheratz
Capella
Zenith
AURIGA
M38
M37
M36
TRIANGULUM
Triangulum Galaxy
COMA BERENICES
LEO MINOR
α Lyn
Castor
Pollux
GEMINI
Elnath
M35
Pleiades
ARIES
Hamal
Algieba
Beehive Cluster
LEO
CANCER
Alhena
Aldebaran
TAURUS
Hyades
Uranus
PISCES
Vindemiatrix
Regulus
CANIS MINOR
Procyon
Betelgeuse
Bellatrix
Menkar
Ecliptic
SEXTANS
ORION
E
VIRGO
MONOCEROS
Orion Nebula
Rigel
W
HYDRA
Sirius
CETUS
CRATER
CANIS MAJOR
ERIDANUS
Adhara
LEPUS
PYXIS
PUPPIS
SE
SW
S
CONSTELLATIONS
Galaxies
Star Clusters
Nebulæ
Planets
Stars
Region of interest
February 1: 11:00 p.m.
February 15: 10:00 p.m.
March 1: 9:00 p.m.

# Highlights in the Southern Sky

**Orion (the Hunter)** is still well placed this month. The Hunter's three belt stars – **Alnitak**, **Alnilam** and **Mintaka** (east to west) – are the most obvious features of this constellation, visible even from light-polluted cities. Just below the belt stars is Orion's sword, and, if you have a keen eye and a dark sky above you, you might catch a faint fuzz in that region. That is the **Great Orion Nebula (Messier 42)**. This is a massive star-forming region of cloud and dust, and it is the closest one to Earth, at 1,500 light-years away, and is roughly 30 light-years across.

There are also wonderful star clusters to enjoy nearby. You can find the magnificent open star cluster **Hyades (Melotte 25)** in the constellation **Taurus (the Bull)**, which can be found just to the northwest of Orion. This cluster contains about 100 stars and is just 150 light-years from Earth.

The brightest star in Taurus is **Aldebaran**, which is clearly visible as an orange-reddish twinkling light in the sky. It looks like it's part of the Hyades cluster, but it's actually about 60 light-years closer. The star is a red giant that lies about 65 light-years away. It's also a variable star, meaning it changes in brightness over time. And, like so many stars, Aldebaran is also part of a binary system.

One of the most beautiful open star clusters is also nearby. The **Pleiades (Messier 45)** – or the Seven Sisters – is 444 light-years from Earth. Though just a handful of stars are visible, it's believed that the entire cluster contains 1,000 stars or more. In Japanese, the name for the cluster is *subaru*. That's right, the Japanese car company is named after this beautiful constellation (hence it's logo, though it contains only six stars).

While **Canis Major (the Great Dog)** is still prominent in the southern sky, **Canis Minor (the Little Dog)**, with its bright white star Procyon, is also visible off Orion's left shoulder. The ancient Greeks named the star **Procyon** because it rose before Kyon, an alternate name for **Sirius**, the Dog Star.

The Hyades (Melotte 25) and the Pleiades (Messier 45)

Procyon
CANIS MINOR
Alphard
HYDRA
Bellatrix
Betelgeuse
ORION
ERIDANUS
Alhena
GEMINI
Pollux
Castor
Beehive Cluster
CANCER
SEXTANS
Hyades
Aldebaran
TAURUS
Elnath
AURIGA
Regulus
Menkar
CETUS
Pleiades
α Lyn
Algieba
LEO
Zenith
Capella
LYNX
Uranus
PERSEUS
Algol
ARIES
Hamal
TRIANGULUM
Mirfak
Kemble's Cascade
NGC 1502
URSA MAJOR
LEO MINOR
Ecliptic
Triangulum Galaxy
Almach
Double Cluster
CAMELOPARDALIS
Merak
Dubhe
PISCES
Mirach
ANDROMEDA
Pacman Nebula
CASSIOPEIA
Polaris
CANES VENATICI
VIRGO
Alioth
W
Andromeda Galaxy
Caph
Kochab
Mizar
COMA BERENICES
E
Alpheratz
CEPHEUS
Coma Cluster
Wizard Nebula
URSA MINOR
Alkaid
Whirlpool Galaxy
δ Cephei
Alderamin
Garnet
PEGASUS
LACERTA
Elephant's Trunk Nebula
DRACO
BOOTES
Cocoon Nebula
CYGNUS
HERCULES
Deneb
Eltanin
NW
NE
N
CONSTELLATIONS
Galaxies
Star Clusters
Nebulæ
Planets
Stars
Region of interest
February 1: 11:00 p.m.
February 15: 10:00 p.m.
March 1: 9:00 p.m.

# Highlights in the Northern Sky

The north is rich with targets this month, though sometimes they get overlooked in favor of the more popular southern sky.

**Cepheus** is an easily identifiable constellation. Looking like a house with a pointed roof, it lies in between **Cassiopeia** and **Ursa Minor (the Little Bear)**.

**Perseus** is found high in the northwest and is home to the much-loved **Double Cluster (NGC 884 and NGC 869)**. From a dark-sky site, the pair can be seen with the naked eye. Grab a pair of binoculars, and the scene is even more spectacular.

The region stretching from Perseus through Cassiopeia and Cepheus offers up a treasure trove of nebulae, including the **Pacman Nebula (NGC 281)**, the **Wizard Nebula (NGC 7380)**, the **Cocoon Nebula (IC 5146)** and the **Elephant's Trunk Nebula (IC 1396A)**. These beautiful clouds of dust and gas make for some stellar photographic targets.

There's also an amazing region that is home to thousands of galaxies, called the **Coma Cluster**. It can be found in the northeast through **Canes Venatici (the Hunting Dogs)**, **Coma Berenices (Berenice's Hair)** and **Virgo (the Maiden)**.

Last but not least, one of the most photographed galaxies is likely the **Whirlpool Galaxy (Messier 51)**, which lies near the first star in the Big Dipper's handle, **Alkaid**.

The Coma Cluster

# March

## March's Events

It's "March madness" of a different kind.

This month is very special to astronomers, as it's the time of year to start running the Messier Marathon. In this celestial treasure hunt, astronomers comb the sky beginning in mid-March to find every one of the 110 Messier objects in the sky in a single night. This year, the full Moon dominates the night sky in the middle of the month, but the sky will be darker in the last week. For most objects, you can use a decent pair of binoculars with a magnification of 10 power or more, and, of course, it's best to get to the darkest location you can. You can find the complete list of Messier objects on pages 114–117.

In the early morning, there's a new trio of planets in the east just before the Sun rises. While Mercury has now gotten too close to the glare of the Sun to be visible, Saturn has joined Venus and Mars. Venus is unmistakable and is roughly 12 degrees above the horizon, with Saturn very low until the end of the month. It's probably best to use binoculars to find it. At the end of the month, Saturn, Venus and Mars begin a little planetary dance low in the southeast before dawn that extends into April. The waning crescent Moon joins them on March 28.

At this time of year, the familiar winter constellations are beginning to set in the west, but that just means the summer constellations are beginning to creep up in the east in the wee hours of the morning.

March 20 is the vernal equinox, an ushering in of spring.

## Calendar of Events

| Day | Time (UTC) | Event |
|---|---|---|
| 2 | 17:35 | New Moon |
| 5 | 12:51 | Jupiter in conjunction with Sun |
| 7 | 06:08 | Uranus 3.7°N of Moon |
| 10 | 10:45 | First quarter of the Moon |
| 10 | 23:05 | Moon at apogee: 404,300 km (251,220 mi.) |
| 12 | 14:00 | Venus 4°N of Mars |
| 13 | | Daylight Saving Time begins (most of U.S. and Canada) |
| 18 | 07:17 | Full Moon |
| 20 | 09:25 | Venus 47°W of Sun (greatest western elongation) |
| 20 | 15:33 | Vernal (spring) equinox |
| 23 | 23:28 | Moon at perigee: 369,800 km (229,783 mi.) |
| 25 | 05:37 | Last quarter of the Moon |
| 28 | 02:54 | Mars 4°N of Moon |

## The Moon This Month

| SUN | MON | TUES | WED | THURS | FRI | SAT |
|---|---|---|---|---|---|---|
| | | 1 | 2<br>New Moon | 3 | 4 | 5 |
| 6 | 7 | 8 | 9 | 10<br>1st Quarter | 11 | 12 |
| 13 | 14 | 15 | 16 | 17 | 18<br>Full Moon | 19 |
| 20 | 21 | 22 | 23 | 24 | 25<br>Last Quarter | 26 |
| 27 | 28 | 29 | 30 | 31 | | |

Once again, the new Moon occurs at the beginning of the month, on March 2.

To spot the planet Uranus, look for it in binoculars on the evening of March 6 above the waxing crescent Moon.

Apogee falls on March 10, which also happens to be the first quarter. The Moon makes a lovely appearance near Pollux, the head of one of the Gemini twins, on the night of March 12.

The full Moon occurs on March 18, with perigee falling on March 23.

The star Pollux, the head of one of the Gemini twins

HERCULES
DRACO
CAMELOPARDALIS
Double Cluster
Andromeda Galaxy
ANDROMEDA
Mizar
Alkaid
Alioth
Dubhe
Almach
Mirfak
Whirlpool Galaxy
Merak
PERSEUS
Triangulum Galaxy
Algol
BOÖTES
URSA MAJOR
CANES VENATICI
LYNX
Capella
TRIANGULUM
Zenith
AURIGA
Izar
Hamal
ARIES
LEO MINOR
α Lyn
Castor
GEMINI
Elnath
Pleiades
Arcturus
COMA BERENICES
Pollux
CANCER
Algieba
Beehive Cluster
Alhena
Aldebaran
TAURUS
Uranus
Ecliptic
Denebola
LEO
CANIS MINOR
Hyades
Vindemiatrix
Regulus
Rosette Nebula
Betelgeuse
Leo Triplet (M65, M66, NGC 3623)
Procyon
Bellatrix
CETUS
Menkar
E
SEXTANS
MONOCEROS
ORION
Orion Nebula
W
VIRGO
CRATER
Alphard
Sirius
Rigel
ERIDANUS
HYDRA
CANIS MAJOR
CORVUS
Spica
Adhara
LEPUS
PYXIS
CONSTELLATIONS
ANTLIA
PUPPIS
COLUMBA
Galaxies
SE
SW
Star Clusters
Nebulæ
S
Planets
Stars
Region of interest
March 1: 11:00 p.m.
March 15: 10:00 p.m. (11:00 p.m. DST)
April 1: 9:00 p.m. (10:00 p.m. DST)

# Highlights in the Southern Sky

The winter constellations are beginning to make their way to the west this month. **Cancer (the Crab)** is high in the southern sky, making the **Beehive Cluster (Messier 44)**, also known as **Praesepe (the Manger)**, a great target. This open star cluster is one of the closest to Earth, lying just 600 light-years away. A glittering celestial jewelry box, it is believed to contain roughly 1,000 stars. In 2012, two exoplanets were discovered orbiting two of the stars in the cluster.

Another wonderful treat is the **Rosette Nebula (NGC 2237)**, which lies in **Monoceros (the Unicorn)** to the right of **Orion (the Hunter)**. It is a beautiful red nebula that is a favorite of astrophotographers and is also home to the open star cluster **NGC 2244**.

**Hydra (the Water Snake)**, the longest of the 88 constellations, winds its way from the southern horizon up toward Cancer.

Another great constellation is **Leo (the Lion)**, which is next to Cancer toward the east. The brightest stars in Leo are **Denebola** and **Regulus**. If you're doing the Messier Marathon, this constellation has several Messier galaxies, though you will need at least a small telescope to view them. The most famous of those objects is the **Leo Triplet**, which consists of three galaxies: **Messier 65**, **Messier 66** and **NGC 3628** (the latter is also known as the **Hamburger Galaxy**).

Leo is also home to **Algieba (gamma Leonis)**, a beautiful orange-yellow double star. It is visible in a small telescope, but you will need high power to separate the double.

The Rosette Nebula (NGC 2237)

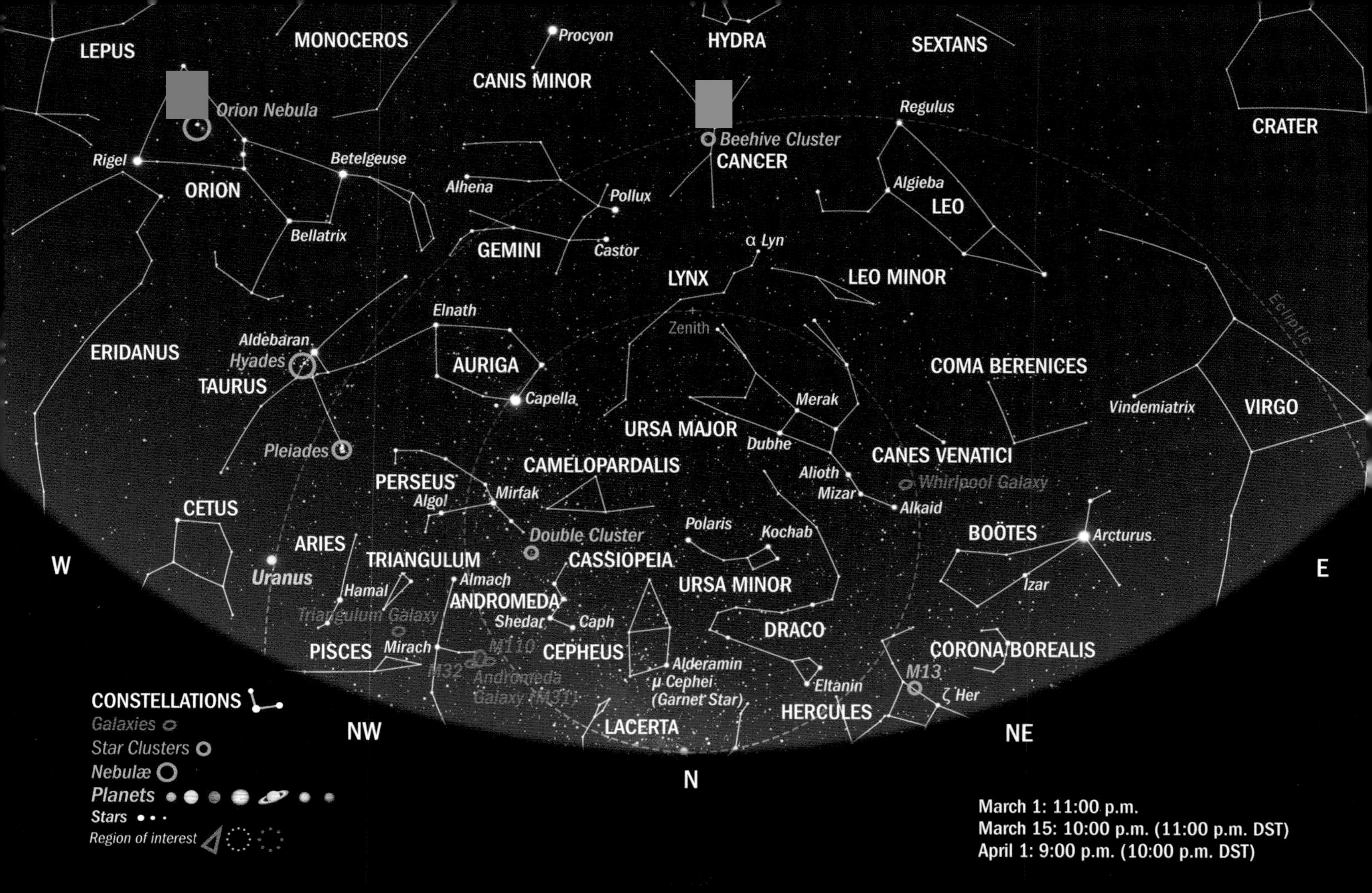
LEPUS
MONOCEROS
Procyon
CANIS MINOR
HYDRA
SEXTANS
CRATER
Orion Nebula
Rigel
ORION
Betelgeuse
Bellatrix
Alhena
GEMINI
Pollux
Castor
Beehive Cluster
CANCER
Regulus
Algieba
LEO
α Lyn
LYNX
LEO MINOR
Ecliptic
ERIDANUS
Aldebaran
Hyades
TAURUS
Elnath
AURIGA
Capella
Zenith
Merak
URSA MAJOR
Dubhe
COMA BERENICES
Vindemiatrix
VIRGO
Pleiades
PERSEUS
Algol
Mirfak
CAMELOPARDALIS
Alioth
Mizar
Alkaid
CANES VENATICI
Whirlpool Galaxy
CETUS
ARIES
TRIANGULUM
Double Cluster
CASSIOPEIA
Polaris
Kochab
URSA MINOR
BOÖTES
Arcturus
Izar
W
E
Uranus
Hamal
Almach
ANDROMEDA
Triangulum Galaxy
Shedar
Caph
DRACO
PISCES
Mirach
M110
CEPHEUS
Alderamin
μ Cephei (Garnet Star)
CORONA BOREALIS
M32
Andromeda Galaxy (M31)
Eltanin
M13
ζ Her
HERCULES
LACERTA
NW
N
NE
CONSTELLATIONS
Galaxies
Star Clusters
Nebulæ
Planets
Stars
Region of interest
March 1: 11:00 p.m.
March 15: 10:00 p.m. (11:00 p.m. DST)
April 1: 9:00 p.m. (10:00 p.m. DST)

# Highlights in the Northern Sky

This month, **Boötes (the Herdsman)** begins to rise low in the northwest. The most prominent star in this kite-like constellation is **Arcturus**, an orange giant star that is 25 times larger than our Sun. It's also older – believed to be about 7.1 billion years old. If you look at Arcturus through a telescope, you will be able to see two stars, as it is a double star. (When you are able to see two stars that are part of a double, it is called "separating" them.)

The constellation starts to rise higher in the east by the end of the month, but if you're looking for Arcturus, the stars of the **Big Dipper's** handle "arc toward Arcturus."

The little house of **Cepheus** still features prominently in the northern sky, with nearby **Cassiopeia**.

If you're going for the Messier Marathon, you'll find three galaxies in the constellation **Andromeda** all clustered together: the **Andromeda Galaxy (Messier 31)**, **Messier 32** and **Messier 110**. The Andromeda Galaxy, which is visible to the unaided eye away from light pollution, is fairly easy to spot as soon as it gets dark. It lies between **Shedar** in Cassiopeia and **Mirach** in the constellation Andromeda. You will need a telescope to see the other two.

The Andromeda Galaxy (Messier 31)

# April

## April's Events

The nights are getting shorter, but, on the plus side, they're also getting warmer.

The morning sky is a cornucopia of planets, with Jupiter, Venus, Mars and Saturn all lining up along the eastern horizon. The four stretch up slightly northward and make for a great pre-sunrise treat beginning on April 15 (as Jupiter will be particularly low on the horizon).

This month there's also a meteor shower, the Lyrids. It's not a strong shower, as it only has a ZHR of roughly 18, but it can produce some fireballs. The peak is on the night of April 22 to 23.

## Calendar of Events

| Day | Time (UTC) | Event |
|---|---|---|
| 1 | 06:24 | New Moon |
| 2 | 23:05 | Mercury at superior conjunction |
| 3 | 17:27 | Uranus 0.6°N of Moon |
| 5 | 01:45 | Mars 0.3°S of Saturn |
| 7 | 19:11 | Moon at apogee: 404,400 km (251,283 mi.) |
| 9 | 06:48 | First quarter of the Moon |
| 16 | 18:55 | Full Moon |
| 19 | 15:16 | Moon at perigee: 365,100 km (226,863 mi.) |
| 22–23 | | Lyrid meteor shower peak |
| 23 | 11:56 | Last quarter of the Moon |
| 24 | 20:56 | Saturn 5°N of Moon |
| 25 | 22:06 | Mars 4°N of Moon |
| 27 | 01:51 | Venus 4°N of Moon |
| 27 | 08:23 | Jupiter 4°N of Moon |
| 29 | 8:09 | Mercury 21°E of Sun (greatest eastern elongation) |
| 30 | 19:56 | Venus 0.2°S of Jupiter |
| 30 | 20:28 | New Moon |
| 30 | 20:42 | Partial solar eclipse (not observable from North America) |

## The Moon This Month

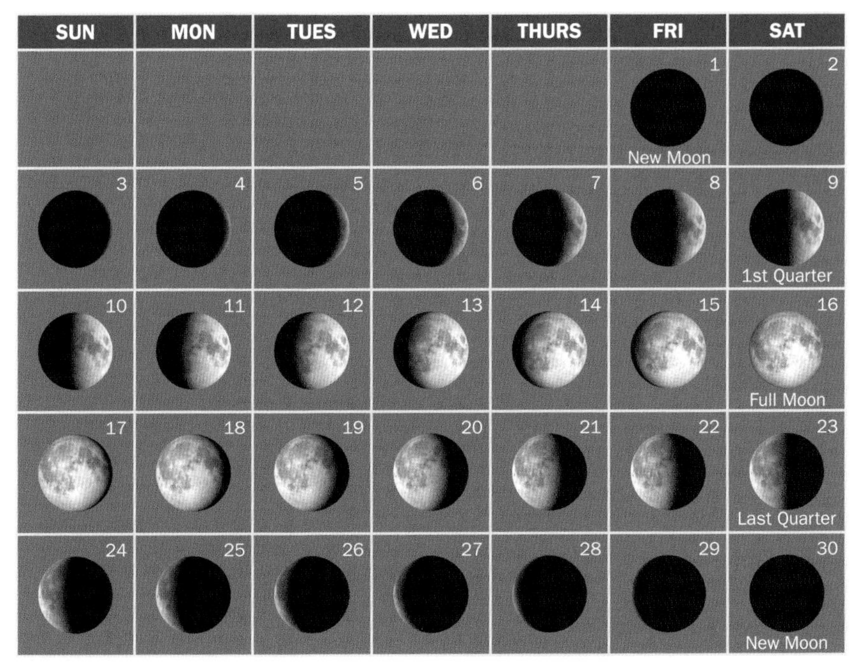


It's no joke: the new Moon falls on April 1 this month and again on April 30. Apogee falls on April 7, while perigee is on April 17.

The evening of April 3 is your last chance to use the Moon to find Uranus before it disappears into the twilight. Look for the planet in binoculars below and to the right of a three-day crescent Moon low in the west.

Between April 24 and April 27, the waning crescent Moon parades by Saturn, Mars, Venus and Jupiter. On April 27, a beautiful crescent Moon sits below and between Jupiter and Venus, the two brightest planets in the night sky.

On April 30, there is a partial solar eclipse that will be visible only in the southern hemisphere (see page 29).

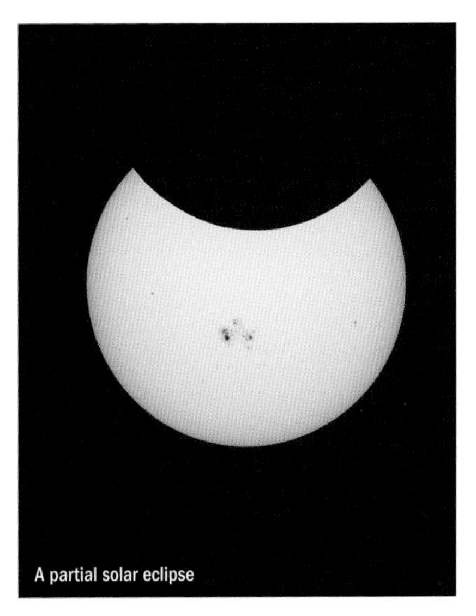
A partial solar eclipse

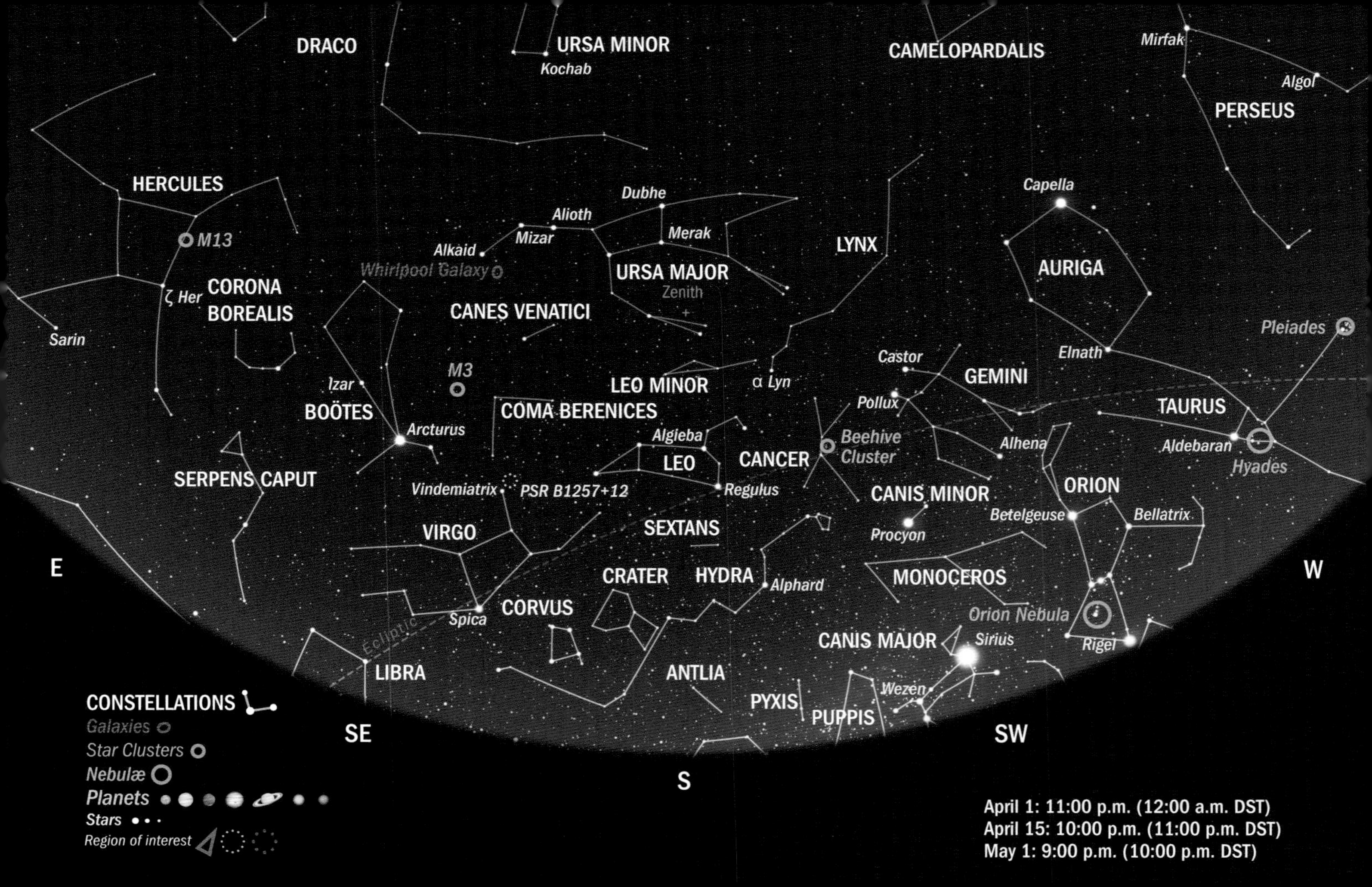
DRACO
URSA MINOR
Kochab
CAMELOPARDALIS
Mirfak
Algol
PERSEUS
HERCULES
M13
Dubhe
Alioth
Mizar
Alkaid
Merak
LYNX
Capella
AURIGA
Whirlpool Galaxy
URSA MAJOR
Zenith
CORONA BOREALIS
ζ Her
Sarin
CANES VENATICI
Pleiades
Elnath
Castor
GEMINI
M3
Izar
LEO MINOR
α Lyn
BOÖTES
COMA BERENICES
Pollux
TAURUS
Arcturus
Algieba
Beehive Cluster
Alhena
Aldebaran
Hyades
LEO
CANCER
SERPENS CAPUT
Vindemiatrix
PSR B1257+12
Regulus
CANIS MINOR
ORION
Betelgeuse
Bellatrix
VIRGO
SEXTANS
Procyon
E
W
CRATER
HYDRA
Alphard
MONOCEROS
CORVUS
Orion Nebula
Spica
Ecliptic
CANIS MAJOR
Sirius
Rigel
LIBRA
ANTLIA
Wezen
PYXIS
PUPPIS
SE
SW
S
CONSTELLATIONS
Galaxies
Star Clusters
Nebulæ
Planets
Stars
Region of interest
April 1: 11:00 p.m. (12:00 a.m. DST)
April 15: 10:00 p.m. (11:00 p.m. DST)
May 1: 9:00 p.m. (10:00 p.m. DST)

# Highlights in the Southern Sky

**Leo (the Lion)** is still high in the east and is a great binocular target, while **Gemini (the Twins)** is now high in the west. The star **Capella** still shines brightly in **Auriga (the Charioteer)** high in the southwest.

**Hydra (the Water Snake)** is still stretching across the southern sky, though it's not a very prominent constellation, despite its length. As in March, the **Beehive Cluster (Messier 44)**, which can be found north of Hydra in the constellation **Cancer (the Crab)**, is in an optimal location for viewing through a pair of binoculars.

One of the largest constellations is now above the horizon: **Virgo (the Maiden)**. It is the second largest of our 88 constellations and is famous for its exoplanets. In 1992, the first exoplanets ever to be discovered were found around pulsar **PSR B1257+12** in Virgo. There have been many more exoplanets found since then. On April 15, a nearly full Moon sits squarely within Virgo, above the constellation's brightest star, **Spica**, which represents an ear of wheat.

Finally, it's time to say goodbye to **Orion (the Hunter)**, the **Hyades (Melotte 25)** and the **Pleiades (Messier 45)** as the wonderful winter trio is hugging the western horizon by the end of the month.

The Beehive Cluster (Messier 44)

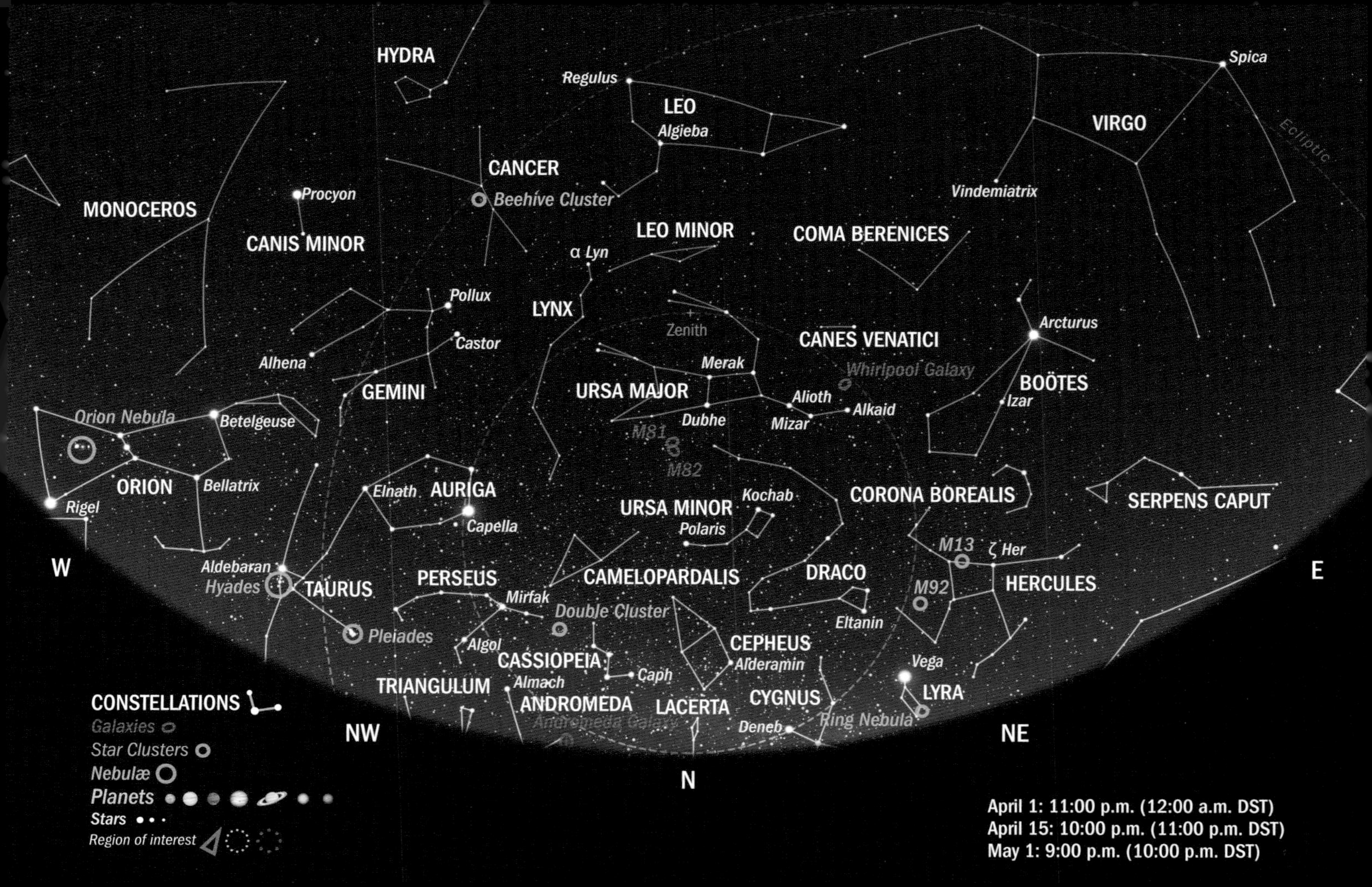
HYDRA
Regulus
LEO
Algieba
VIRGO
Spica
Ecliptic
CANCER
Beehive Cluster
Vindemiatrix
MONOCEROS
Procyon
CANIS MINOR
LEO MINOR
COMA BERENICES
α Lyn
Pollux
LYNX
Zenith
Arcturus
Castor
CANES VENATICI
Alhena
Merak
Whirlpool Galaxy
GEMINI
URSA MAJOR
BOÖTES
Alioth
Izar
Alkaid
Orion Nebula
Betelgeuse
Dubhe
Mizar
M81
M82
ORION
Bellatrix
Elnath
AURIGA
Kochab
CORONA BOREALIS
Rigel
Capella
URSA MINOR
SERPENS CAPUT
Polaris
M13
ζ Her
W
Aldebaran
DRACO
E
Hyades
TAURUS
PERSEUS
CAMELOPARDALIS
Mirfak
M92
HERCULES
Double Cluster
Pleiades
Eltanin
Algol
CEPHEUS
CASSIOPEIA
Alderamin
Vega
Caph
TRIANGULUM
Almach
LYRA
CONSTELLATIONS
ANDROMEDA
LACERTA
CYGNUS
Andromeda Galaxy
Deneb
Ring Nebula
Galaxies
NW
NE
Star Clusters
Nebulæ
N
Planets
Stars
Region of interest
April 1: 11:00 p.m. (12:00 a.m. DST)
April 15: 10:00 p.m. (11:00 p.m. DST)
May 1: 9:00 p.m. (10:00 p.m. DST)

## Highlights in the Northern Sky

**Ursa Major (the Great Bear)** and its asterism the **Big Dipper** are high in the north, though upside down.

This month is a great time to check out **Messier 81 (Bode's Galaxy)** and **Messier 82 (Cigar Galaxy)** in Ursa Major. They can be seen in dark-sky sites and are in the same field of view through binoculars or a small telescope.

By mid-month, **Hercules** has risen above the northeastern horizon. The constellation is home to one of the most stunning globular clusters in the night sky, **Messier 13**. It's so magnificent that in dark-sky locations it can be seen as a faint fuzz with the naked eye. With binoculars or a small telescope, the collection of more than several hundred thousand stars is truly a sight to behold. The cluster is roughly 25,000 light-years from Earth.

There's another globular cluster in Hercules, though not quite as stunning: **Messier 92**. It is also visible with the naked eye, though only under good dark-sky conditions.

In between Hercules and **Boötes (the Herdsman)**, you will find the small arc of stars called **Corona Borealis (The Northern Crown)**. To the Mi'kmaq First Nation, these stars represent the winter den of Muin, the She-Bear.

Messier 82 (Cigar Galaxy)

# May

## May's Events

More comfortable nights are now here. Since the nights are also getting shorter, this is a good time to start planning for viewing the stars (and more) from summertime retreats, whether it's time away at the cottage or just camping under a curtain of stars.

Our fabulous foursome of planets – Venus, Jupiter, Mars and Saturn – are still well placed in the morning sky. Venus is now sinking lower toward the horizon, however. After sunset on May 2, look for the planet Mercury low in the west, about 5 degrees west of a young crescent Moon. On May 29, Jupiter and Mars are in conjunction less than half a degree apart low in the eastern dawn sky. Bring out the binoculars or a small telescope for that one.

There's also a meteor shower this month, the Eta Aquariids, which began on April 19 but peaks on the night of May 4 to 5. Though it's better seen from the southern hemisphere, the shower could produce 10 to 30 meteors an hour.

## Calendar of Events

| Day | Time (UTC) | Event |
|---|---|---|
| 2 | 14:17 | Mercury 1.8°N of Moon |
| 4–5 | | Eta Aquariid meteor shower peak |
| 5 | 12:46 | Moon at apogee: 405,300 km (251,842 mi.) |
| 9 | 00:21 | First quarter of the Moon |
| 16 | 04:11 | Total lunar eclipse |
| 16 | 04:14 | Full Moon |
| 17 | 15:23 | Moon at perigee: 360,300 km (223,880 mi.) |
| 21 | 19:14 | Mercury at inferior conjunction |
| 22 | 04:43 | Saturn 4°N of Moon |
| 22 | 18:43 | Last quarter of the Moon |
| 24 | 19:24 | Mars 3°N of Moon |
| 24 | 23:59 | Jupiter 3°N of Moon |
| 27 | 02:52 | Venus 0.2°N of Moon |
| 29 | 08:57 | Mars 0.6°S of Jupiter |
| 30 | 11:30 | New Moon |

## The Moon This Month

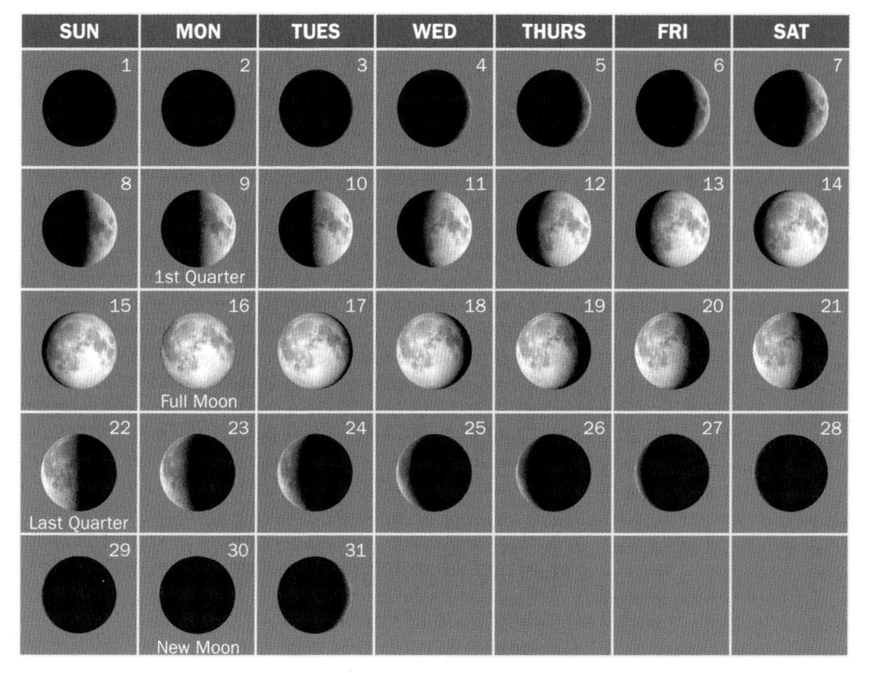


This is the first month of the year that doesn't start with a new Moon. Apogee falls on May 5, and perigee falls on May 17, the day after the full Moon.

On May 16, there will be a total lunar eclipse that will be visible either in part or in its entirety across North America. (See page 27.)

The Moon swings through the planets in the early morning starting on May 22, when it will be roughly 4 degrees south of Saturn. On May 25, a crescent Moon lies below Jupiter, roughly 4 degrees away from it, and Mars sits less than 3 degrees west of Jupiter. Finally, on May 27, a thin crescent Moon – only about 8 percent illuminated – will lie just to the southeast of the brightest planet in our Solar System, Venus, low on the horizon.

Seeing multiple planets pairing with the Moon is an enchanting experience, like this Moon-Venus-Mercury combo in the dawn sky.

CYGNUS
DRACO
URSA MINOR
Kochab
Capella
AURIGA
Eltanin
LYNX
Dubhe
Vega
LYRA
Ring Nebula
Alioth
Merak
Mizar
HERCULES
Alkaid
URSA MAJOR
M13
Castor
GEMINI
Zenith
CORONA
BOREALIS
CANES
VENATICI
LEO MINOR
α Lyn
ζ Her
Pollux
Sarin
BOÖTES
Izar
COMA BERENICES
Beehive
Cluster
Algieba
Alhena
Arcturus
M53
LEO
CANCER
Rasalhague
Regulus
ORION
Vindemiatrix
SERPENS CAPUT
Virgo Galaxy Cluster
CANIS MINOR
VIRGO
SEXTANS
Procyon
E
MONOCEROS
W
Sombrero Galaxy
OPHIUCHUS
CRATER
HYDRA
Alphard
Spica
CORVUS
LIBRA
Sabik
Ecliptic
ANTLIA
CONSTELLATIONS
SCORPIUS
Menkent
LUPUS
Galaxies
SE
SW
Star Clusters
Nebulæ
S
Planets
Stars
Region of interest
May 1: 11:00 p.m. (12:00 a.m. DST)
May 15: 10:00 p.m. (11:00 p.m. DST)
June 1: 9:00 p.m. (10:00 p.m. DST)

## Highlights in the Southern Sky

Directly to the south are two fairly faint constellations, **Corvus (the Crow)** and **Crater (the Cup)**. The pair lie low in the south, directly above **Hydra (the Water Snake)**, which is fully visible (though faint).

**Leo (the Lion)** and **Virgo (the Maiden)** are the most noticeable constellations in the south. Virgo's brightest star, **Spica**, shines brightly. Spica is a variable star, meaning its brightness changes over time, and it lies roughly 262 light-years away from Earth. It's also the 14th brightest star in the sky and part of a binary system.

In Virgo lies the **Virgo Galaxy Cluster**, home to roughly 2,000 galaxies. In between Virgo and Corvus is the **Sombrero Galaxy (Messier 104)**. We see this galaxy edge-on, with its prominent dust lane best viewed through large telescopes. It lies about 28 million light-years away and is one of the most massive objects in the Virgo Galaxy Cluster, weighing in at 800 million Suns.

Above Virgo is the constellation **Coma Berenices (Berenice's Hair)**. Within this constellation is **Messier 53**, a beautiful, tight globular cluster that is one of the most distant globular clusters, at 59,700 light-years away. The cluster can be seen through small telescopes, though larger ones will reveal the individual stars themselves.

The Sombrero Galaxy (Messier 104)

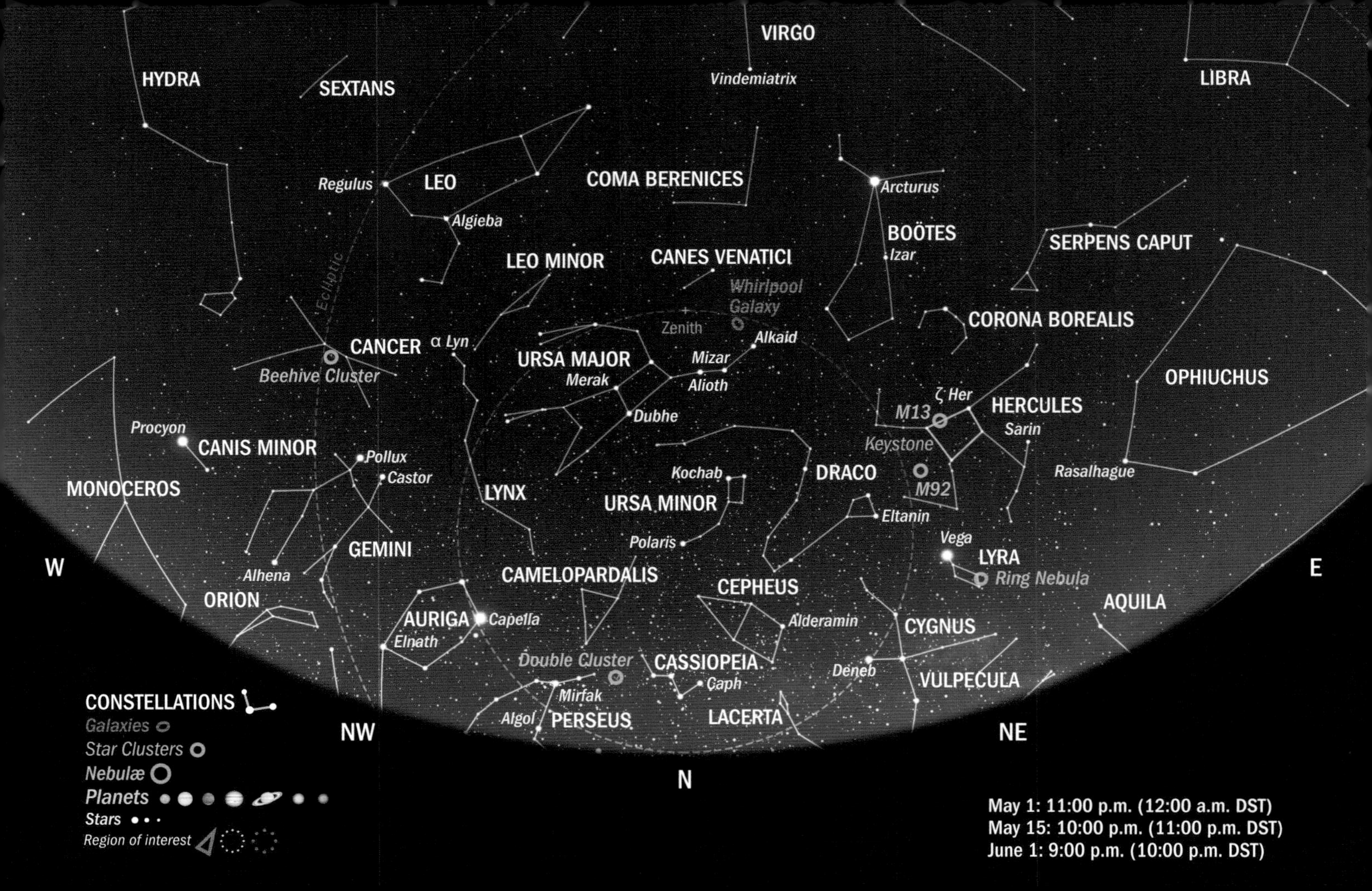
HYDRA
SEXTANS
VIRGO
Vindemiatrix
LIBRA
Regulus
LEO
COMA BERENICES
Arcturus
Algieba
BOÖTES
Izar
SERPENS CAPUT
LEO MINOR
CANES VENATICI
Whirlpool Galaxy
Ecliptic
Zenith
CORONA BOREALIS
CANCER
α Lyn
Alkaid
URSA MAJOR
Mizar
Merak
Alioth
OPHIUCHUS
Beehive Cluster
ζ Her
HERCULES
M13
Dubhe
Sarin
Procyon
CANIS MINOR
Keystone
Pollux
Castor
Kochab
DRACO
Rasalhague
MONOCEROS
M92
LYNX
URSA MINOR
Eltanin
Vega
Polaris
GEMINI
LYRA
W
CAMELOPARDALIS
Ring Nebula
E
Alhena
CEPHEUS
AQUILA
ORION
AURIGA
Capella
Alderamin
CYGNUS
Elnath
Double Cluster
CASSIOPEIA
Deneb
VULPECULA
Caph
Mirfak
CONSTELLATIONS
Algol
PERSEUS
LACERTA
Galaxies
NW
NE
Star Clusters
Nebulæ
N
Planets
Stars
Region of interest
May 1: 11:00 p.m. (12:00 a.m. DST)
May 15: 10:00 p.m. (11:00 p.m. DST)
June 1: 9:00 p.m. (10:00 p.m. DST)

# Highlights in the Northern Sky

**Cygnus (the Swan)** is rising in the northeast, along with **Lyra (the Lyre)**. The brightest star in the latter constellation is **Vega**. It is the fifth brightest star and is just 25 light-years from Earth. In popular culture, it was featured as the location of an alien signal in Carl Sagan's book *Contact* (and the movie of the same name).

The constellation is also home to the **Ring Nebula (Messier 57)**. The nebula looks like what its name suggests, a ring, and is about 2,000 light-years away. Though not visible to the unaided eye, it is visible through small telescopes.

The mighty **Hercules** is now firmly placed in the northeast, presenting a great chance to check out the globular cluster **Messier 13**. There's another great nearby globular cluster, **Messier 92**, which is one of the brightest in our galaxy, with about 300,000 tightly packed stars. You can find it to the west of Hercules and its asterism the **Keystone**.

You might notice a bright star now rising higher in the northeast. That's **Deneb**, the brightest star in Cygnus, representing the tail of the swan. The star is roughly 200 times more massive than our Sun and is one of the most distant stars we can see with the naked eye, lying some 1,500 light-years away.

The **Big Dipper** asterism is high above, with the star at the tip of the handle, **Alkaid**, lying almost at the zenith.

**Cepheus** has moved farther to the northeast, while **Cassiopeia** lies low in the north and **Perseus** sinks lower in the northwest. **Auriga (the Charioteer)** is now beginning to disappear in the west.

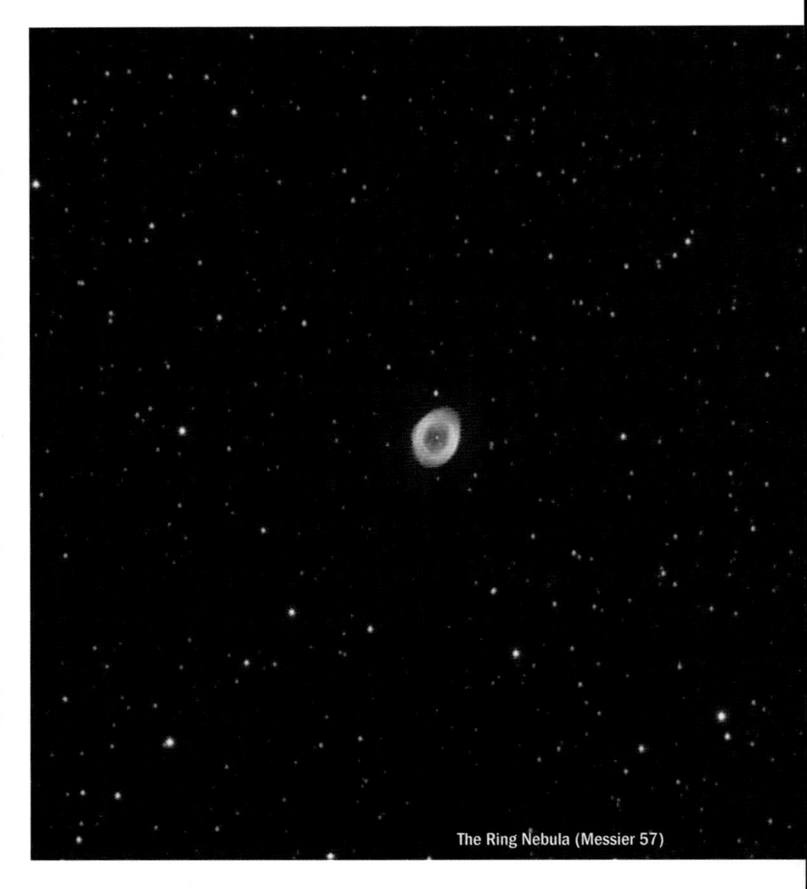

The Ring Nebula (Messier 57)

# June

## June's Events

The constellations Scorpius and Sagittarius begin to rise in the June sky

Summer nights may be short, but there is certainly no shortage of targets to take in. The summer sky holds many amazing nebulae, open clusters and globular clusters as the richest part of the Milky Way stretches across the sky. You'll find a wealth of targets particularly in the Sagittarius and Scorpius constellations.

Mercury returns to the morning sky, low on the horizon. It reaches greatest western elongation on June 16. Look for it low in the east before sunrise, about 10 degrees below and to the left of brilliant Venus. Jupiter and Mars are now fairly high up as they begin their journey as nighttime objects.

We ring in the summer with the solstice on June 21.

## Calendar of Events

| Day | Time (UTC) | Event |
|---|---|---|
| 2 | 01:14 | Moon at apogee: 406,200 km (252,401 mi.) |
| 7 | 14:48 | First quarter of the Moon |
| 14 | 11:52 | Full Moon |
| 14 | 23:21 | Moon at perigee: 357,400 km (222,078 mi.) |
| 16 | 14:55 | Mercury 23°W of Sun (greatest western elongation) |
| 18 | 12:22 | Saturn 4°N of Moon |
| 21 | 03:11 | Last quarter of the Moon |
| 21 | 09:14 | Summer solstice |
| 21 | 13:31 | Jupiter 3°N of Moon |
| 22 | 18:16 | Mars 0.9°N of Moon |
| 26 | 08:11 | Venus 3°S of Moon |
| 27 | 08:18 | Mercury 4°S of Moon |
| 29 | 02:52 | New Moon |
| 29 | 06:08 | Moon at apogee: 406,600 km (252,650 mi.) |

## The Moon This Month

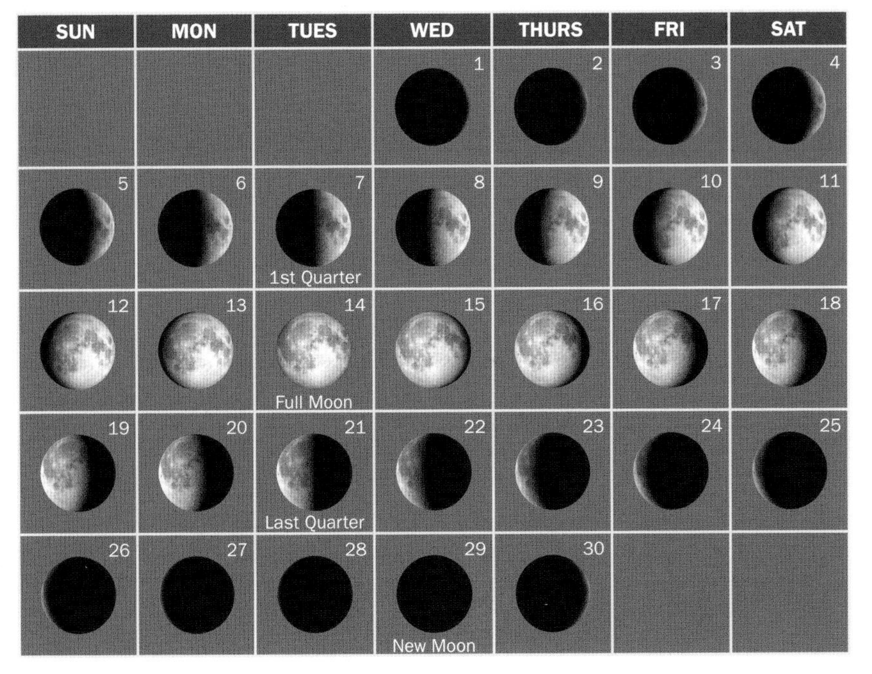


The month starts off with the Moon at apogee on June 2 and once again on June 29. Interestingly, perigee falls on June 14 along with the full Moon and the second apogee corresponds with the new Moon. On the days following June 14, seaside dwellers will experience large tides.

On June 26, challenge yourself to see if you can find a very thin crescent Moon lying about 3 degrees away from Venus in the early morning eastern sky. On the next day, for a bigger challenge, try to observe the Moon about 4 degrees away from Mercury, closer to the horizon.

A crescent Moon pairs with Venus near the end of June

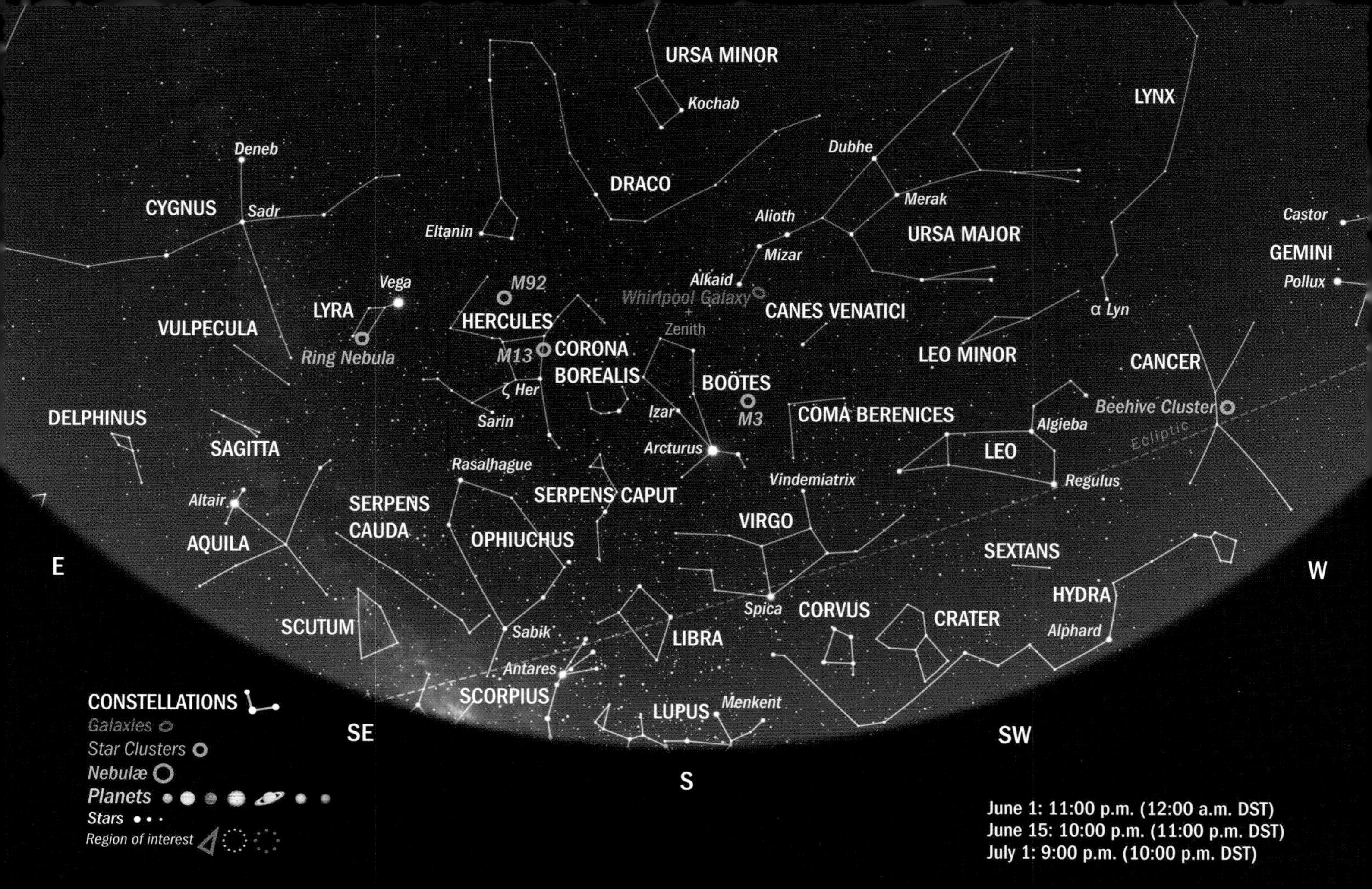
URSA MINOR
Kochab
LYNX
Deneb
DRACO
Dubhe
CYGNUS
Sadr
Merak
Castor
Alioth
URSA MAJOR
Eltanin
GEMINI
Mizar
Vega
M92
Alkaid
Pollux
LYRA
Whirlpool Galaxy
CANES VENATICI
α Lyn
HERCULES
VULPECULA
Zenith
Ring Nebula
M13
CORONA BOREALIS
LEO MINOR
CANCER
BOÖTES
ζ Her
Beehive Cluster
DELPHINUS
Izar
COMA BERENICES
Sarin
M3
Algieba
Ecliptic
SAGITTA
Arcturus
LEO
Rasalhague
Vindemiatrix
Regulus
Altair
SERPENS CAPUT
SERPENS CAUDA
VIRGO
OPHIUCHUS
SEXTANS
AQUILA
E
W
HYDRA
Spica
CORVUS
CRATER
SCUTUM
Sabik
Alphard
LIBRA
Antares
SCORPIUS
Menkent
LUPUS
CONSTELLATIONS
Galaxies
Star Clusters
Nebulæ
Planets
Stars
Region of interest
SE
S
SW
June 1: 11:00 p.m. (12:00 a.m. DST)
June 15: 10:00 p.m. (11:00 p.m. DST)
July 1: 9:00 p.m. (10:00 p.m. DST)

## Highlights in the Southern Sky

**Libra (the Scales)** is in a nice position in the southern sky. Just above it is **Boötes (the Herdsman)**, with its star **Arcturus** shining brightly. Boötes's position also provides a great chance to observe the globular cluster **Messier 3** in the nearby **Canes Venatici (the Hunting Dogs)**.

You may notice a reddish star rising in the southeast. That's **Antares**, a massive red supergiant similar to Orion's Betelgeuse. The star is several hundred times the diameter of the Sun and 10,000 times more luminous. It is also a binary star, though its much dimmer companion is invisible. The star gets its name from a Greek word meaning "rival to Ares (Mars)," due to its reddish color. Eventually, like Betelgeuse, it will end its life in a spectacular supernova explosion.

The constellation **Ophiuchus (the Serpent Bearer)** is now high in the east. Its brightest star, **Rasalhague**, marks the tip of the constellation.

**Hercules** is now positioned high in the east, providing a great opportunity to catch the globular clusters **Messier 13** and **Messier 92**.

The center of the Milky Way with bright red Antares in the constellation Scorpius

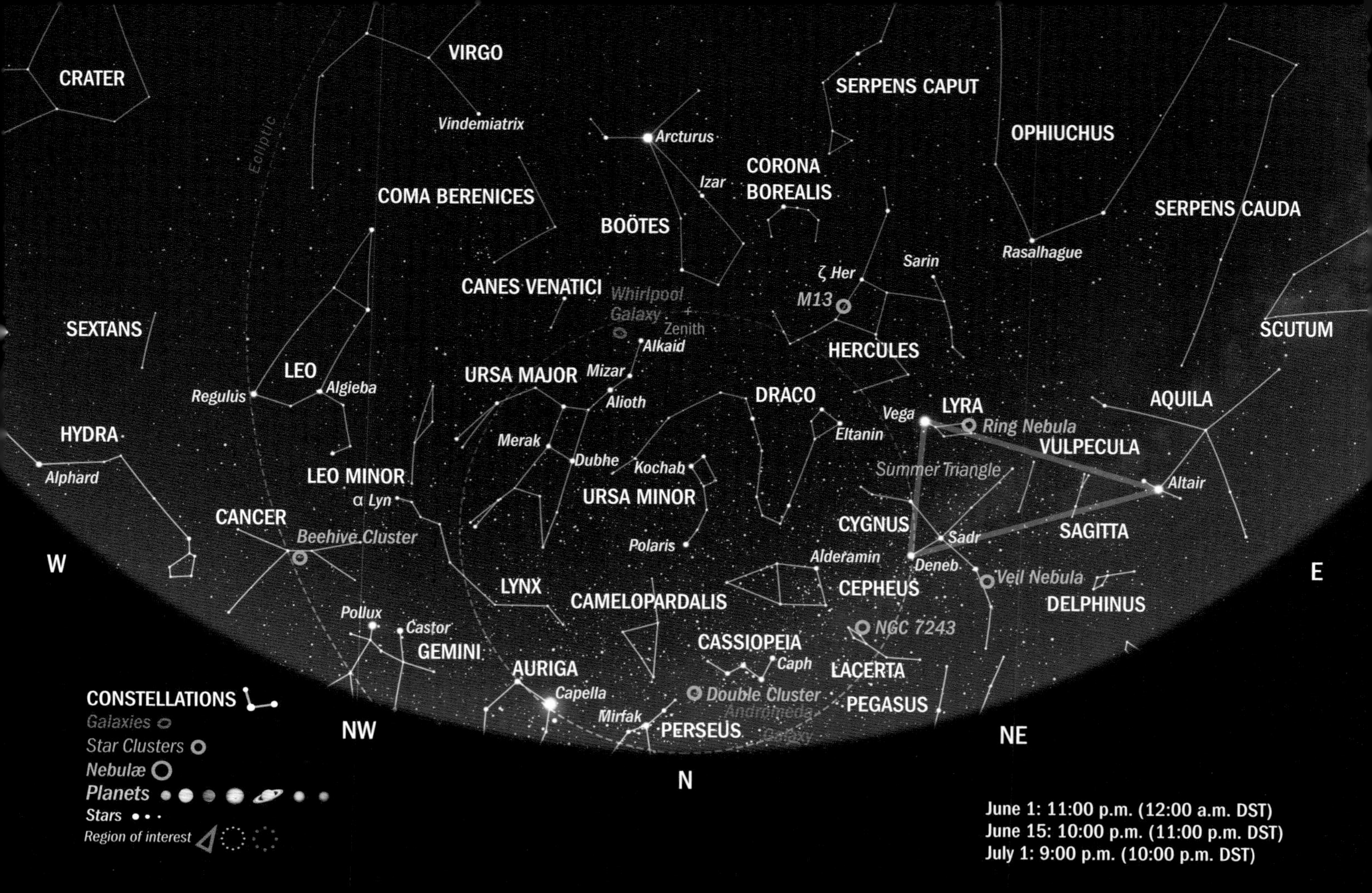
CRATER
VIRGO
Vindemiatrix
Arcturus
SERPENS CAPUT
OPHIUCHUS
CORONA BOREALIS
Izar
COMA BERENICES
BOÖTES
SERPENS CAUDA
Rasalhague
Sarin
ζ Her
CANES VENATICI
Whirlpool Galaxy
Zenith
M13
SEXTANS
SCUTUM
Alkaid
HERCULES
LEO
URSA MAJOR
Mizar
Regulus
Algieba
Alioth
DRACO
LYRA
AQUILA
Vega
Ring Nebula
HYDRA
Merak
Eltanin
VULPECULA
Dubhe
Kochab
Summer Triangle
Alphard
LEO MINOR
Altair
URSA MINOR
α Lyn
CYGNUS
CANCER
Sadr
SAGITTA
Beehive Cluster
Polaris
Alderamin
W
Deneb
E
Veil Nebula
LYNX
CAMELOPARDALIS
CEPHEUS
DELPHINUS
Pollux
Castor
NGC 7243
CASSIOPEIA
GEMINI
Caph
AURIGA
LACERTA
Double Cluster
CONSTELLATIONS
Capella
Andromeda Galaxy
PEGASUS
Galaxies
Mirfak
NW
PERSEUS
NE
Star Clusters
Nebulæ
N
Planets
Stars
Region of interest
June 1: 11:00 p.m. (12:00 a.m. DST)
June 15: 10:00 p.m. (11:00 p.m. DST)
July 1: 9:00 p.m. (10:00 p.m. DST)

# Highlights in the Northern Sky

**Cassiopeia** is now low in the north, with the house-like **Cepheus** sitting on its side below **Draco (the Dragon)**. Draco winds between **Ursa Major (the Great Bear)** and **Hercules**.

The **Summer Triangle** is also starting to rise higher in the sky. This asterism is made up of three stars: **Altair** in the constellation **Aquila (the Eagle)**, **Vega** in **Lyra (the Lyre)** and **Deneb** in **Cygnus (the Swan)**. These three constellations continue to move higher up in the sky over the months, all the way into October. Cygnus is a treasure trove of beautiful clusters, and if you grab a pair of binoculars, you can get lost in this rich region of the **Milky Way**.

To the northeast is **Lacerta (the Lizard)**, a small, kite-like constellation that is situated between **Cassiopeia** and Cygnus. A nice open cluster, **NGC 7243** lies near the head of Lacerta, which you can see through binoculars or small telescopes.

June is also a perfect month to see **noctilucent clouds**. These clouds form high in the atmosphere at altitudes of around 76 to 85 kilometers (47 to 53 miles) and occur in the Northern Hemisphere beginning around the middle of May. They are not completely understood, but it's believed they are caused by meteoroid dust in the mesosphere. These iridescent clouds used to only be seen in polar regions, but they have more recently been spotted farther south, even as far south as Los Angeles.

Noctilucent clouds

June 2022

# July

## July's Events

The Milky Way is now beautifully placed in the southern sky. Sagittarius and Scorpius are low in the south, and there are numerous targets to enjoy with the naked eye, binoculars or a telescope. There are nebulae and clusters galore.

Venus is still shining brightly in the eastern morning sky, while Mars, Jupiter and Saturn are now nighttime targets, so long as you're willing to stay up late.

By July 15, Saturn begins to rise after 10:00 p.m. local time, sitting in Capricornus (though on July 15 there is an almost full Moon that will hamper the view), while Jupiter rises after midnight in Pisces.

Earth is at aphelion on July 4, when it is the farthest from the Sun in its orbit. (Contrary to what some may believe, Earth isn't farther from the Sun in January.)

## Calendar of Events

| Day | Time (UTC) | Event |
|---|---|---|
| 4 | 02:59 | Earth at aphelion: 1.0167 AU |
| 7 | 02:14 | First quarter of the Moon |
| 13 | 09:08 | Moon at perigee: 357,300 km (222,016 mi.) |
| 13 | 18:37 | Full Moon |
| 15 | 20:16 | Saturn 4°N of Moon |
| 16 | 19:30 | Mercury at superior conjunction |
| 18 | 00:50 | Neptune 4°N of Moon |
| 19 | 00:55 | Jupiter 2°N of Moon |
| 20 | 14:18 | Last quarter of the Moon |
| 21 | 16:46 | Mars 3°W Moon |
| 26 | 10:22 | Moon at apogee: 406,300 km (252,463 mi.) |
| 26 | 14:12 | Venus 4°S of Moon |
| 28 | 17:55 | New Moon |

## The Moon This Month

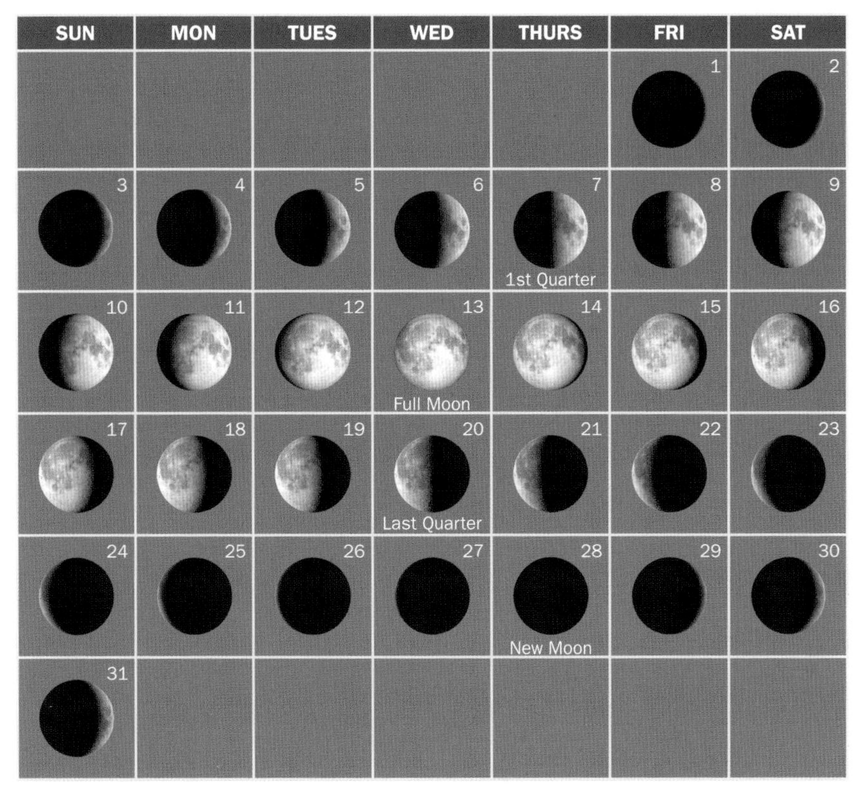


This month, perigee and the full Moon once again occur on the same day, July 13. Seaside dwellers will experience large tides in the days following. Apogee falls on July 26, with the new Moon falling on July 28.

On July 10, a waxing gibbous Moon will be roughly 2 degrees from Antares in Scorpius. On July 12, the nearly full Moon will sit squarely in Sagittarius and will be 99.2 percent illuminated.

On the evening of July 17, look for Neptune in the same binocular field as the waning gibbous Moon, 4 degrees above. A few nights later, on July 22, look for Uranus less than 1 degree north of the waning crescent Moon at the crack of dawn. You can use binoculars or zoom in with a telescope to see the pair.

At dawn on July 21, the waning crescent Moon and Mars will be about 3 degrees apart. Keen-eyed larks will enjoy a sliver of the Moon roughly 3 degrees from Venus in the early morning sky on July 26. The pair will be hugging the horizon in Gemini.

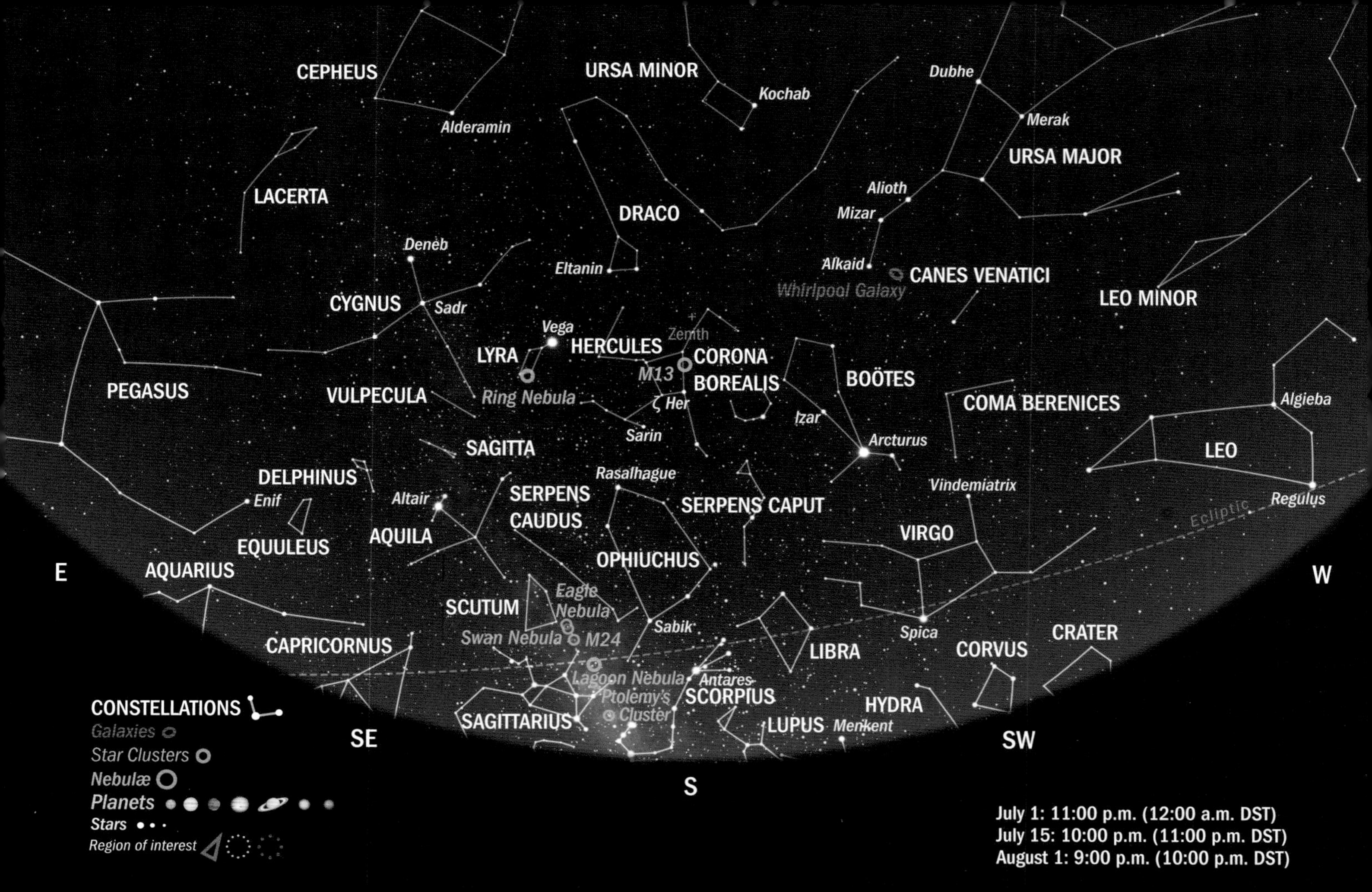
CEPHEUS
URSA MINOR
Kochab
Dubhe
Merak
URSA MAJOR
Alderamin
LACERTA
DRACO
Alioth
Mizar
Deneb
Eltanin
Alkaid
CANES VENATICI
Whirlpool Galaxy
LEO MINOR
CYGNUS
Sadr
Vega
Zenith
LYRA
HERCULES
CORONA BOREALIS
M13
BOÖTES
PEGASUS
VULPECULA
Ring Nebula
ζ Her
COMA BERENICES
Izar
Algieba
Sarin
Arcturus
LEO
SAGITTA
DELPHINUS
Rasalhague
Vindemiatrix
Regulus
Enif
Altair
SERPENS CAUDUS
SERPENS CAPUT
Ecliptic
VIRGO
EQUULEUS
AQUILA
OPHIUCHUS
E
AQUARIUS
W
Eagle Nebula
SCUTUM
Sabik
Swan Nebula
M24
LIBRA
Spica
CRATER
CAPRICORNUS
CORVUS
Lagoon Nebula
Antares
SCORPIUS
Ptolemy's Cluster
HYDRA
SAGITTARIUS
LUPUS
Menkent
SE
SW
S
CONSTELLATIONS
Galaxies
Star Clusters
Nebulæ
Planets
Stars
Region of interest
July 1: 11:00 p.m. (12:00 a.m. DST)
July 15: 10:00 p.m. (11:00 p.m. DST)
August 1: 9:00 p.m. (10:00 p.m. DST)

# Highlights in the Southern Sky

There's never a shortage of beautiful things to see in the night sky, but the summertime sky holds some of the best – and most easily accessible – sights.

First, we start with **Sagittarius (the Archer)**, which is well positioned in the southern sky. This constellation contains numerous naked-eye targets and is an absolute thrill to view through binoculars. Sagittarius is easy to spot as it looks like a teapot.

The most notable and easy-to-spot feature is the **Sagittarius Star Cloud (Messier 24)**, found just north of the teapot's lid. This cloud is located 10,000 light-years from Earth and is about 600 light-years wide. You can't miss it in dark sky locations: It's the fuzzy patch in the sky that looks like... a cloud. Through binoculars it's even more astounding, even from a light-polluted city.

The **Swan Nebula (Messier 17)**, also known as the Omega Nebula, can be found in Sagittarius, above the "lid" of the teapot. This nebula is clearly visible through small telescopes.

Above that lies the **Eagle Nebula (Messier 16)**. This is the subject of the Hubble Space Telescope's famous photograph *The Pillars of Creation*. It is visible through large telescopes and is a favorite target of many astrophotographers.

The **Lagoon Nebula (Messier 8)** is a wonderful emission nebula also found in Sagittarius. Though faint, it can be seen unaided under dark-sky conditions. It even has its own star cluster, **NGC 6530**, which is clearly visible through binoculars.

Then there's the constellation **Scorpius (the Scorpion)**, which has plenty of beautiful star clusters of its own, including **Ptolemy's Cluster (Messier 7)**, found between Scorpius and Sagittarius. It was named after the Greek astronomer and mathematician Claudius Ptolemy, who first recorded it in the 2nd century CE.

The Eagle Nebula (Messier 16)

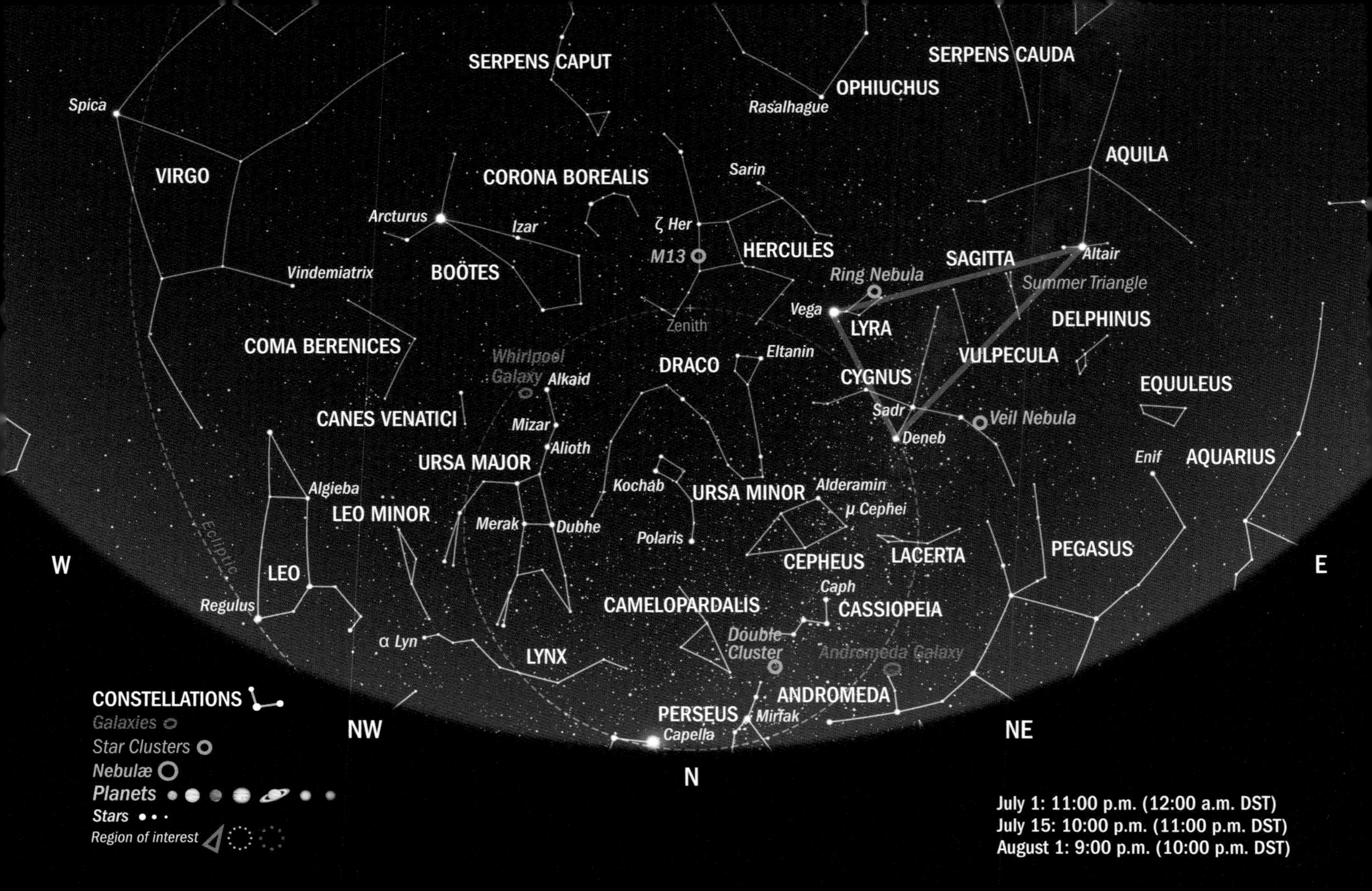
SERPENS CAPUT
SERPENS CAUDA
OPHIUCHUS
Rasalhague
Spica
VIRGO
CORONA BOREALIS
Sarin
AQUILA
Arcturus
Izar
ζ Her
M13
HERCULES
SAGITTA
Altair
Vindemiatrix
BOÖTES
Ring Nebula
Summer Triangle
Vega
LYRA
DELPHINUS
Zenith
COMA BERENICES
Whirlpool Galaxy
Alkaid
DRACO
Eltanin
CYGNUS
VULPECULA
EQUULEUS
CANES VENATICI
Mizar
Sadr
Veil Nebula
Alioth
Deneb
URSA MAJOR
Enif
AQUARIUS
Algieba
Kochab
URSA MINOR
Alderamin
LEO MINOR
μ Cephei
Merak
Dubhe
Polaris
PEGASUS
W
LEO
Ecliptic
CEPHEUS
LACERTA
E
Caph
Regulus
CAMELOPARDALIS
CASSIOPEIA
Double Cluster
Andromeda Galaxy
α Lyn
LYNX
ANDROMEDA
CONSTELLATIONS
Galaxies
PERSEUS
Mirfak
NW
Capella
NE
Star Clusters
Nebulæ
N
Planets
July 1: 11:00 p.m. (12:00 a.m. DST)
Stars
July 15: 10:00 p.m. (11:00 p.m. DST)
Region of interest
August 1: 9:00 p.m. (10:00 p.m. DST)

# Highlights in the Northern Sky

**Ursa Major (the Great Bear)**, with its hard-to-miss asterism the **Big Dipper**, is now beginning to sink toward the northwest, and **Cepheus** can be found in the northeast as it begins to look like a tipped-over house.

**Cygnus (the Swan)**, another rich part of the Milky Way, is well-placed high in the northeast, and the **Summer Triangle** – with the stars **Deneb**, **Vega** and **Altair** – is easy to make out.

Cygnus has a spectacular nebula in its midst. The **Veil Nebula (NGC 6960, 6992 and 6995)** is a remnant of an ancient supernova that lies 2,100 light-years away. It is found south of **Sadr**, the center star in the "cross" of the constellation. The Veil Nebula is actually made up of two parts: the Western Veil Nebula and the Eastern Veil Nebula. Altogether, it measures 10 light-years across, the equivalent of six Moons lined side by side. Viewing these nebulae requires dark skies and large-aperture telescopes.

**Pegasus**, with its distinct square shape, is beginning to rise in the northeast along with **Andromeda**.

If you find yourself in a dark-sky location, try to spot tiny **Delphinus (the Dolphin)** in the east and **Lacerta (the Lizard)**, which lies between Cepheus and **Hercules**.

The Summer Triangle asterism shines in the sky

# August

## August's Events

The daylight hours are now beginning to get noticeably shorter, which is good news for astronomers. Though summer is starting to wind down, there's still plenty of seasonal treats to enjoy.

The most anticipated meteor shower of the year, the Perseids, takes center stage this month. The shower can produce upward of 120 meteors per hour on the peak night, which falls on August 12 to 13. It's the perfect time of year to enjoy it, as the nights are warmer and the skies tend to be clearer. The only downside to the shower this year is that the full Moon falls during the peak, so only the brightest meteors will be visible. Fortunately, the Perseids rarely disappoint and can produce some stunning fireballs.

## Calendar of Events

| Day | Time (UTC) | Event |
|---|---|---|
| 5 | 11:06 | First quarter of the Moon |
| 10 | 17:14 | Moon at perigee: 359,800 km (223,569 mi.) |
| 12 | 01:36 | Full Moon |
| 12 | 03:55 | Saturn 4°N of Moon |
| 12–13 | | Perseid meteor shower peak |
| 14 | 16:35 | Saturn at opposition |
| 15 | 09:37 | Jupiter 1.9°N of Moon |
| 19 | 04:36 | Last quarter of the Moon |
| 19 | 12:16 | Mars 3°S of Moon |
| 22 | 21:53 | Moon at apogee: 405,400 km (251,904 mi.) |
| 25 | 20:58 | Venus 4°S of Moon |
| 27 | 08:17 | New Moon |
| 27 | 16:14 | Mercury 27°E of Sun (greatest eastern elongation)) |
| 29 | 18:51 | Mercury 5°S of Moon |

## The Moon This Month

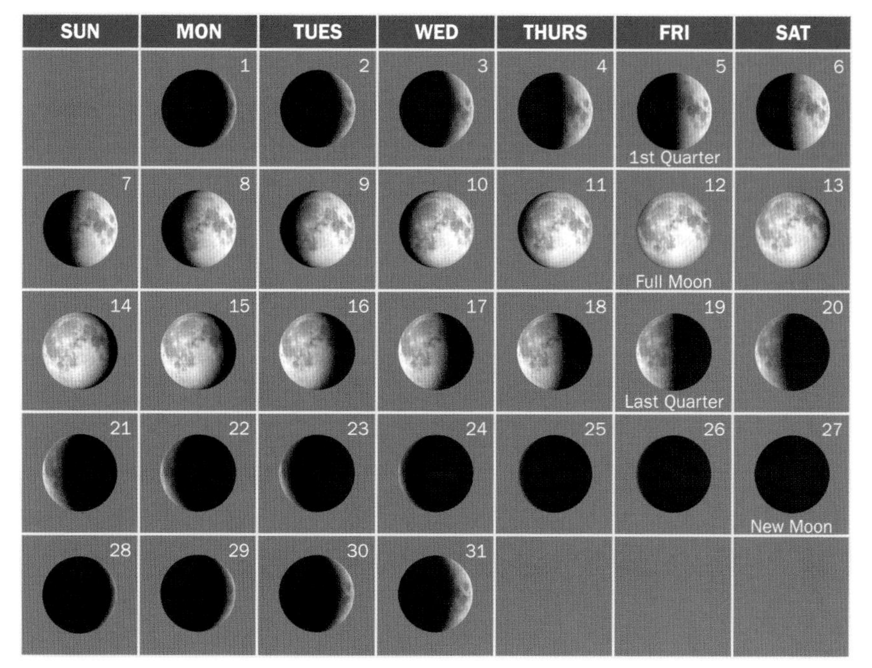

The Perseids streak across the sky in mid-August

The Moon is at perigee on August 10, with the full Moon occurring just two days later, on the night of August 11 to 12. The full Moon is also about 5 degrees from Saturn on that night. On August 15, the Moon sits less than 3 degrees from Jupiter. On August 19, the Moon lies less than 2 degrees north of Mars.

Apogee occurs on August 22 and the new Moon on August 27.

Finally, the Moon will be barely a sliver on August 25, illuminated just 3.5 percent, but it will be about 6 degrees above a brightly shining Venus near the horizon just before sunrise. If you are up for a challenge, look for Mercury 5 degrees below a thin crescent Moon after sunset on August 29.

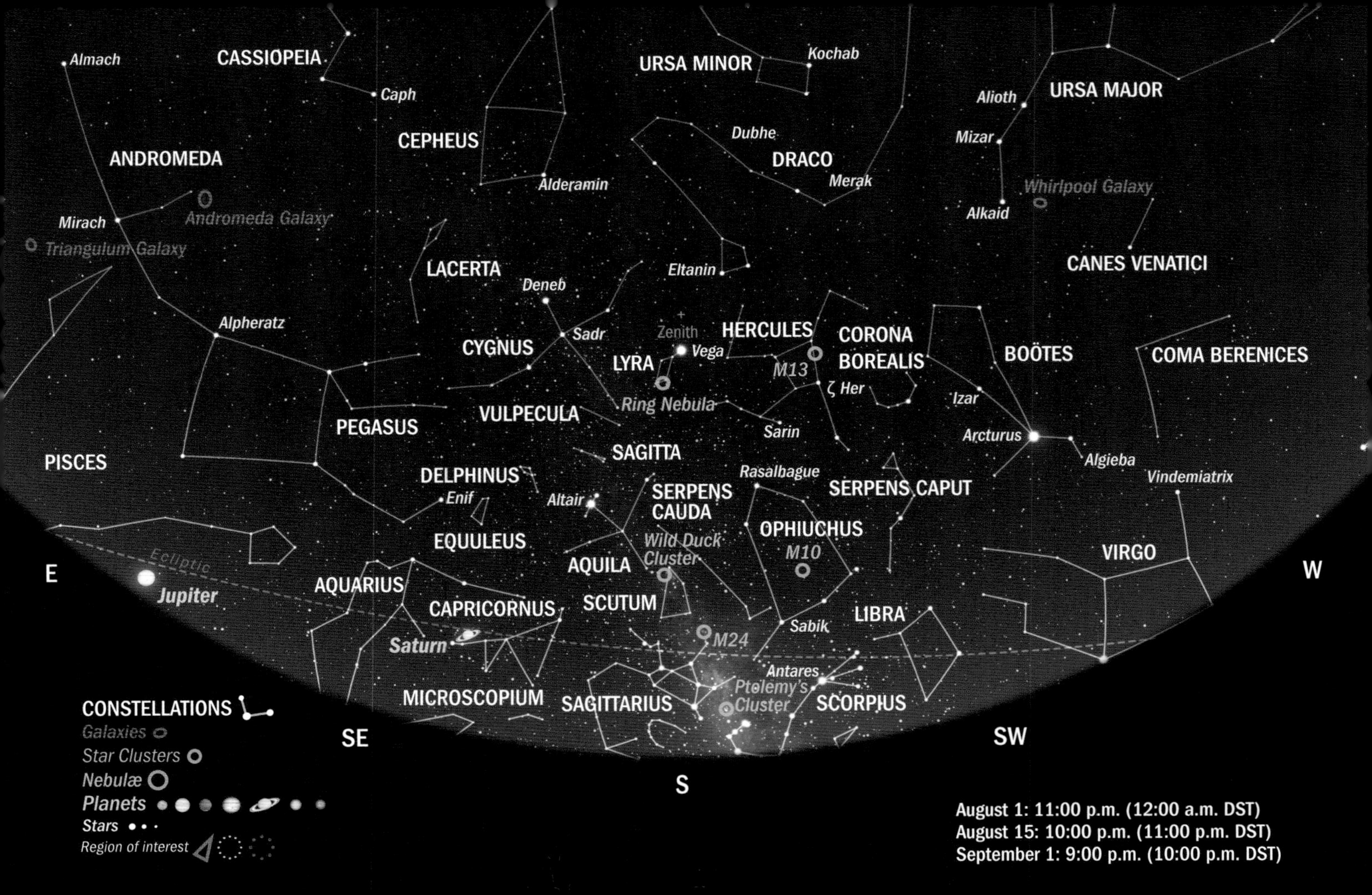
Almach
CASSIOPEIA
Caph
URSA MINOR
Kochab
URSA MAJOR
Alioth
Mizar
Alkaid
Whirlpool Galaxy
CANES VENATICI
CEPHEUS
ANDROMEDA
Alderamin
Dubhe
DRACO
Merak
Mirach
Andromeda Galaxy
Triangulum Galaxy
LACERTA
Deneb
Eltanin
Alpheratz
CYGNUS
Sadr
Zenith
Vega
LYRA
HERCULES
M13
CORONA BOREALIS
ζ Her
BOÖTES
COMA BERENICES
Izar
Arcturus
Algieba
Vindemiatrix
Ring Nebula
Sarin
PEGASUS
VULPECULA
SAGITTA
PISCES
DELPHINUS
Enif
Altair
Rasalbague
SERPENS CAUDA
SERPENS CAPUT
EQUULEUS
OPHIUCHUS
Wild Duck Cluster
M10
AQUILA
VIRGO
E
Ecliptic
Jupiter
AQUARIUS
CAPRICORNUS
SCUTUM
LIBRA
W
Sabik
Saturn
M24
Antares
Ptolemy's Cluster
MICROSCOPIUM
SAGITTARIUS
SCORPIUS
SE
SW
S
CONSTELLATIONS
Galaxies
Star Clusters
Nebulæ
Planets
Stars
Region of interest
August 1: 11:00 p.m. (12:00 a.m. DST)
August 15: 10:00 p.m. (11:00 p.m. DST)
September 1: 9:00 p.m. (10:00 p.m. DST)

# Highlights in the Southern Sky

The "super star" constellations of the southern sky, **Sagittarius (the Archer)** and **Scorpius (the Scorpion)**, are still well placed to be enjoyed.

**Ophiuchus (the Serpent Bearer)** sits above Scorpius, so it is in a good place to observe the globular cluster **Messier 10**. You can spot Messier 10 almost at Ophiuchus's center.

One of the great things about the summer is that observers can enjoy the richness of the Milky Way from the north all the way to the south. A pair of binoculars reveals a mass of stars from **Cassiopeia**, **Cygnus (the Swan)**, **Aquila (the Eagle)** and **Scutum (the Shield)** right through to Sagittarius. Whether you're on a dock at a cottage, outside your tent or even in a city park, there's just so much to enjoy.

You can still find **Saturn** in **Capricornus (the Sea Goat)**. Though you can make out a pancake-like shape through binoculars, the best view of our Solar System's most famous ringed planet is through telescopes big or small. Jupiter rises just after 10:00 p.m. local time in the east. If you're willing to stay up late, Mars begins to grace the night sky, rising in the east after midnight.

The constellation of Scutum can be found just above Sagittarius, and near its tip is the **Wild Duck Cluster (Messier 11)**. This open cluster of stars is located roughly 6,200 light-years from Earth and is one of the most densely populated open clusters known, containing approximately 2,900 stars.

The globular cluster Messier 10

AQUILA
Saturn
OPHIUCHUS
Altair
Rasalhague
SAGITTA
DELPHINUS
Summer Triangle
SERPENS
LIBRA
EQUULEUS
Sarin
Ring Nebula
VULPECULA
Enif
HERCULES
LYRA
CYGNUS
ζ Her
Vega
AQUARIUS
CORONA BOREALIS
M13
Sadr
Zenith
Deneb
Eltanin
Izar
PEGASUS
Arcturus
DRACO
LACERTA
BOÖTES
Alderamin
μ Cephei
VIRGO
URSA MINOR
CEPHEUS
Alkaid
Whirlpool Galaxy
Mizar
Kochab
Alpheratz
Vindemiatrix
Alioth
Caph
Polaris
Andromeda Galaxy
Ecliptic
W
COMA BERENICES
CANES VENATICI
CASSIOPEIA
Jupiter
E
Dubhe
ANDROMEDA
Mirach
PISCES
Merak
URSA MAJOR
Double Cluster
LEO
Triangulum Galaxy
Almach
LEO MINOR
CAMELOPARDALIS
Mirfak
TRIANGULUM
PERSEUS
Algol
LYNX
Capella
NW
AURIGA
NE
N
CONSTELLATIONS
Galaxies
Star Clusters
Nebulæ
Planets
Stars
Region of interest
August 1: 11:00 p.m. (12:00 a.m. DST)
August 15: 10:00 p.m. (11:00 p.m. DST)
September 1: 9:00 p.m. (10:00 p.m. DST)

# Highlights in the Northern Sky

**Ursa Minor (the Little Bear)** hangs upside down in the northern sky this month, while **Ursa Major (the Great Bear)** is found in the northwest with the bowl of the **Big Dipper** at the ready to scoop up its smaller bear brother.

**Draco (the Dragon)** is also high in the north, winding its way between the two bears toward the mighty **Hercules**.

The "W" of **Cassiopeia** sits in the northeast, and its right-hand downward point can guide you to the **Andromeda Galaxy**. It is truly an amazing galaxy to enjoy during the summer. This beautiful spiral is 2.4 million light-years from Earth and has a diameter of 200,000 light-years. It is on a collision course with the Milky Way, but there's no need to panic: This event won't happen for another 4.5 billion years.

**Cepheus** is now high in the northeast. This is also a great time to find **Mu (μ) Cephei**, or "the Garnet Star," so named for its deep reddish color. It is a variable star that fluctuates in brightness over time. In the sky chart for this month, Mu Cephei is the star under the "μ." Look for a ruddy star though binoculars starting at Alderamin to get your bearings.

**Pegasus** is now clear in the east, marked by its "Great Square," and **Delphinus (the Dolphin)** lies between Pegasus and **Aquila (the Eagle)**.

**Perseus** is now rising in the northeast with its brightest star **Mirfak**, which means the **Double Cluster (NGC 869 and NGC 884)** is rising along with it. These two clusters are a favorite for summertime star parties.

**Arcturus**, the brightest star in **Boötes (the Herdsman)**, is now in the west. By the end of the month, it begins to disappear below the horizon.

The constellation Cassiopeia

# September

## September's Events

The nights are definitely longer now, which is both good news and bad news. For one, it means the amazing targets we've enjoyed over the summer months are now beginning to set earlier.

Sagittarius is still visible with its beautiful clusters and nebulae, but you'll only have until the end of the month to enjoy it, and you'll need a low horizon.

Saturn is still well-placed in Cancer, with Jupiter in Pisces.

Mars rises close to 11:30 p.m. local time in the east. If you're willing to stay up late, look for two bright red objects in the east after 1:00 a.m. local time. One is Mars and the other to the west is Aldebaran, the brightest star in Taurus. The pair will be about 3 degrees apart.

Finally, we welcome the fall season on September 23 with the autumnal equinox.

The Sagittarius Star Cloud (Messier 24)

## Calendar of Events

| Day | Time (UTC) | Event |
|---|---|---|
| 3 | 18:08 | First quarter of the Moon |
| 7 | 18:17 | Moon at perigee: 364,500 km (226,490 mi.) |
| 8 | 10:31 | Saturn 4°N of Moon |
| 10 | 09:59 | Full Moon |
| 11 | 15:11 | Jupiter 1.8°N of Moon |
| 17 | 01:41 | Mars 4°S of Moon |
| 17 | 21:52 | Last quarter of the Moon |
| 19 | 14:44 | Moon at apogee: 404,600 km (251,407 mi.) |
| 23 | 01:04 | Autumnal (fall) equinox |
| 23 | 06:47 | Mercury at inferior conjunction |
| 25 | 21:54 | New Moon |

## The Moon This Month

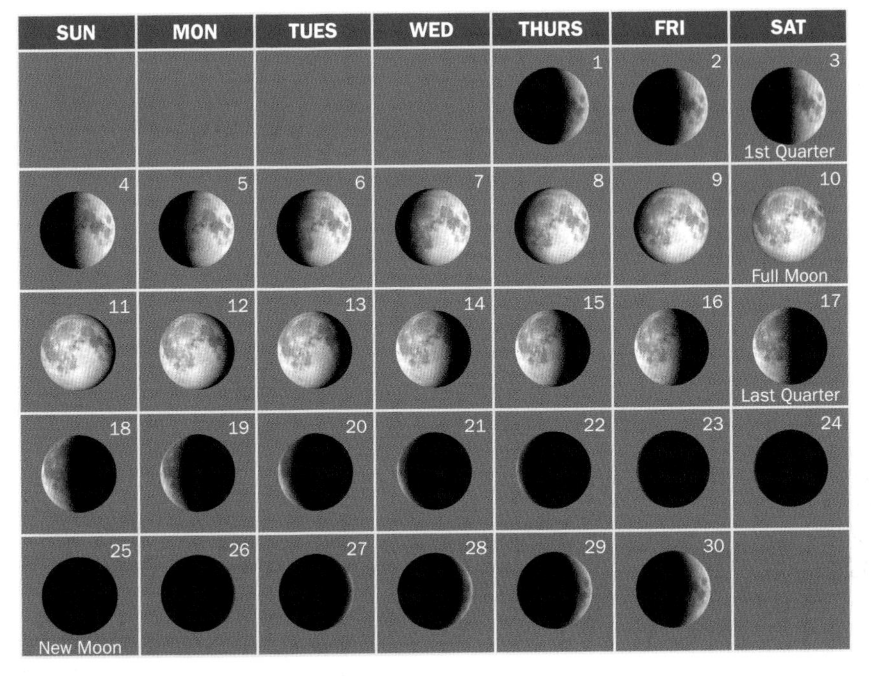

The full Moon meets Jupiter this month

On September 3, the Moon will be about 5 degrees from Antares, found in Scorpius, though both will be low in the southwest just after sunset. The Moon will be just past the first quarter and should present a beautiful sight.

Perigee falls on September 7 and the full Moon on September 10. Apogee occurs on September 19.

On September 10 and 11, the full Moon will be near Jupiter.

And finally, the last quarter Moon will be about 3 degrees from Mars and 10 degrees from Aldebaran on September 17.

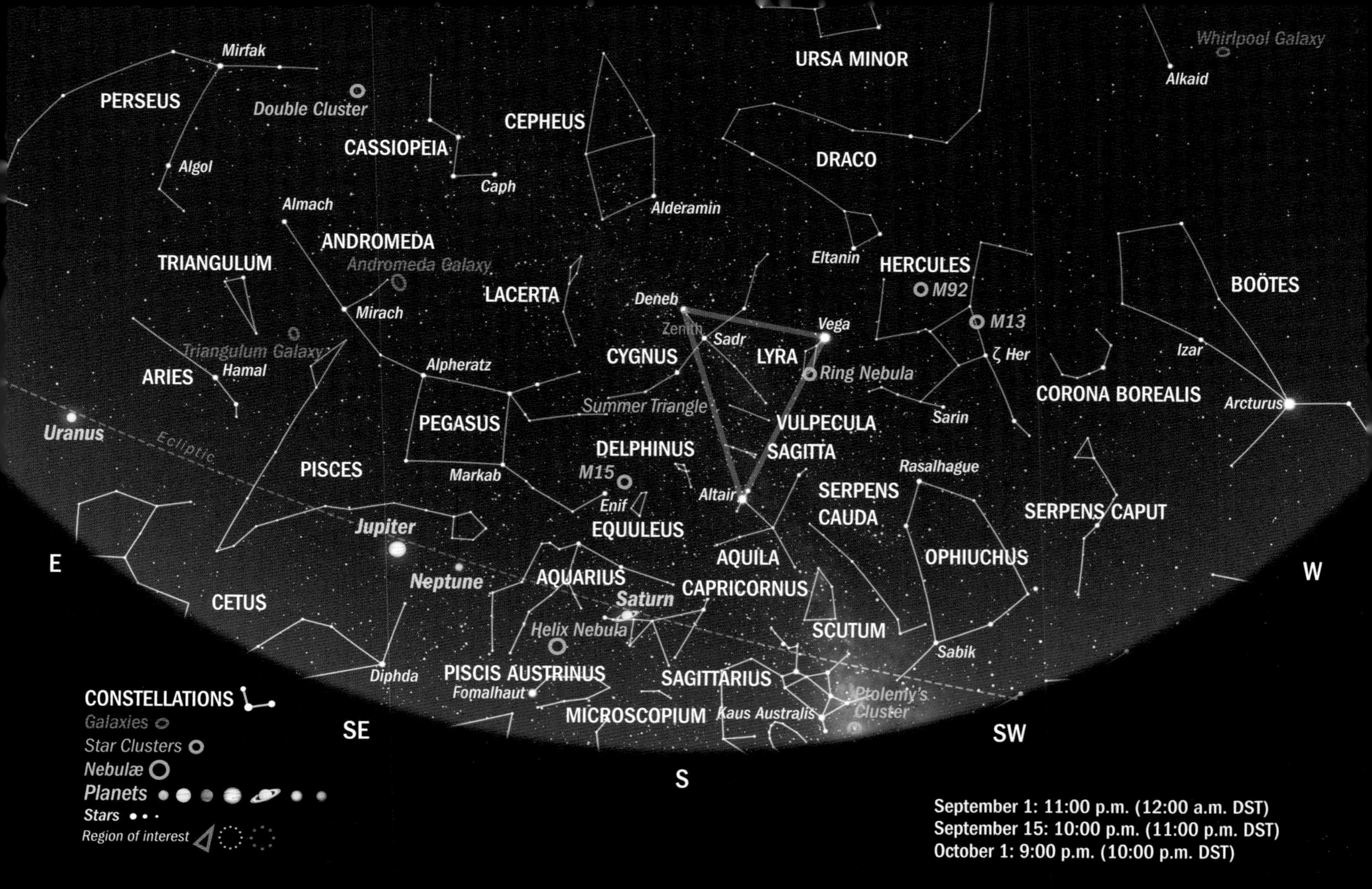
URSA MINOR
Whirlpool Galaxy
Alkaid
Mirfak
PERSEUS
Double Cluster
CEPHEUS
CASSIOPEIA
DRACO
Algol
Caph
Alderamin
Almach
ANDROMEDA
Andromeda Galaxy
TRIANGULUM
Eltanin
HERCULES
M92
BOÖTES
LACERTA
Deneb
Mirach
Zenith
Sadr
Vega
M13
Triangulum Galaxy
CYGNUS
LYRA
ζ Her
Izar
Hamal
Alpheratz
Ring Nebula
ARIES
CORONA BOREALIS
Arcturus
Summer Triangle
Sarin
Uranus
PEGASUS
VULPECULA
Ecliptic
DELPHINUS
SAGITTA
PISCES
Markab
M15
Rasalhague
Altair
SERPENS
CAUDA
Enif
SERPENS CAPUT
Jupiter
EQUULEUS
AQUILA
OPHIUCHUS
E
W
Neptune
AQUARIUS
CAPRICORNUS
CETUS
Saturn
SCUTUM
Helix Nebula
Sabik
Diphda
PISCIS AUSTRINUS
SAGITTARIUS
Fomalhaut
Ptolemy's Cluster
MICROSCOPIUM
Kaus Australis
SE
SW
S
CONSTELLATIONS
Galaxies
Star Clusters
Nebulæ
Planets
Stars
Region of interest
September 1: 11:00 p.m. (12:00 a.m. DST)
September 15: 10:00 p.m. (11:00 p.m. DST)
October 1: 9:00 p.m. (10:00 p.m. DST)

## Highlights in the Southern Sky

The constellation **Sculptor** lies low in the southern sky, along with **Piscis Austrinus (the Southern Fish)**. Above it is **Capricornus (the Sea Goat)**, where **Saturn** sits this month.

**Aquarius (the Water Bearer)** is now high in the southeast. An interesting target near the constellation is the **Helix Nebula (NGC 7293)**, sometimes referred to as the Eye of God. This planetary nebula formed after a star similar to our own Sun died and shed its outer shell of gas. What's left behind is a white dwarf star. You can see it in dark-sky locations by finding **Fomalhaut** in the constellation Piscis Austrinus and then looking northward by 10 degrees.

**Jupiter** sits between **Pisces (the Fishes)** and **Cetus (the Whale)** this month. One of the interesting things observers can do is look at the planet over several nights in a row. A pair of binoculars will reveal four of the planet's moons: **Galileo**, **Europa**, **Io** and **Callisto**.

The **Summer Triangle** now lies high in the south as it makes its way to the southwest. This is your last chance to observe the globular cluster **Messier 10** as **Ophiuchus (the Serpent Bearer)** begins to set in the southwest.

You can find three small constellations in the south: **Sagitta (the Arrow)**, **Equuleus (the Little Horse)** and **Delphinus (the Dolphin)**. Farther west of Sagitta is a wonderful asterism called the **Coathanger (Collinder 399)**, which is best seen through binoculars or a small telescope. It looks like an upside-down coat hanger and is difficult to miss. Look for it on the lowest side of the Summer Triangle, one-third of the way from **Altair** to **Vega**.

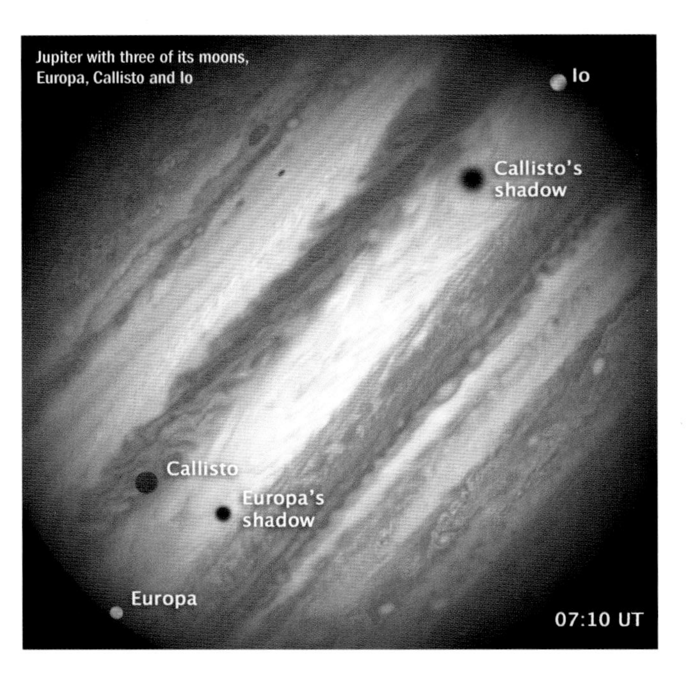

Jupiter with three of its moons, Europa, Callisto and Io

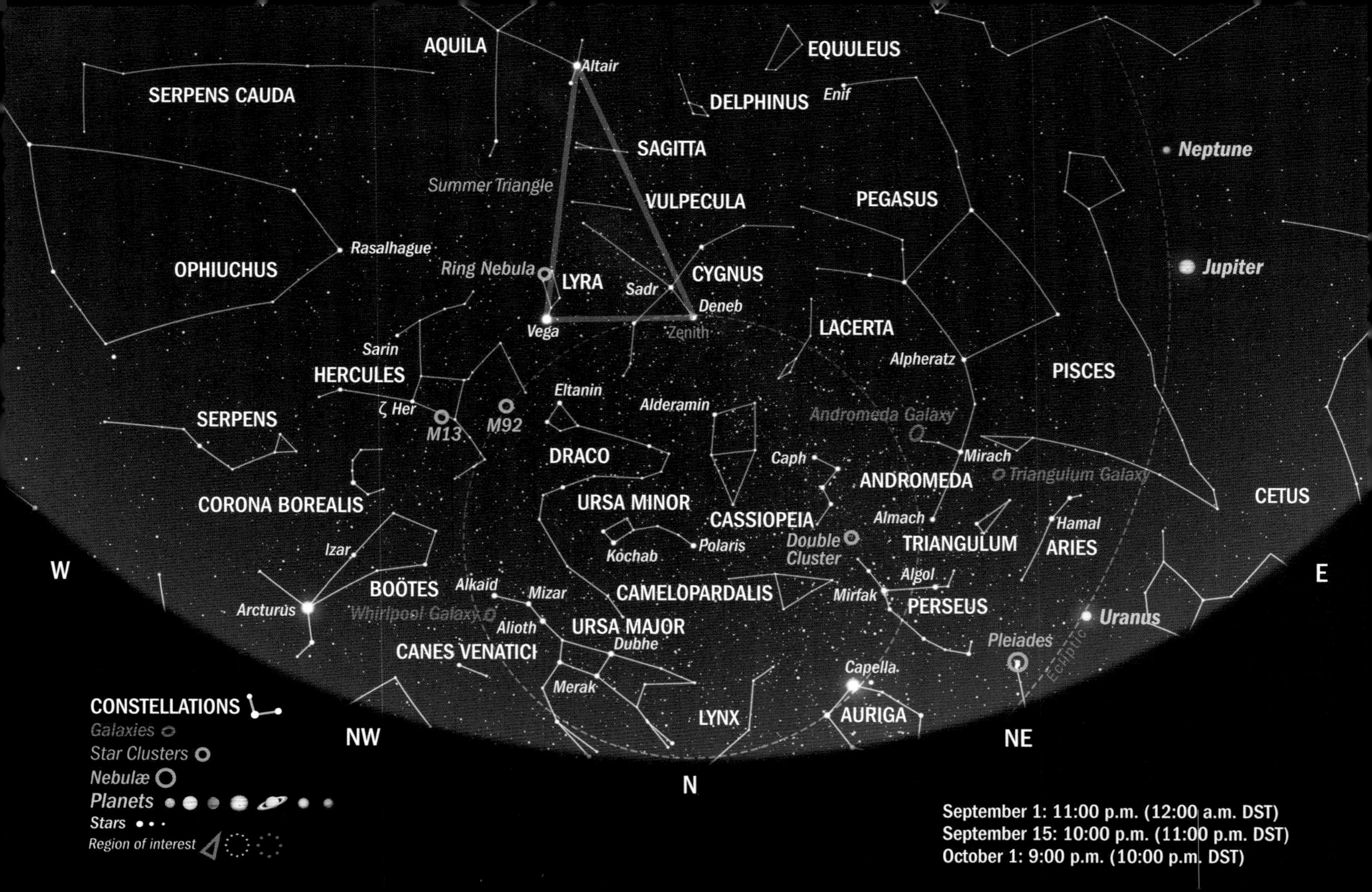

AQUILA
Altair
EQUULEUS
SERPENS CAUDA
DELPHINUS
Enif
SAGITTA
Neptune
Summer Triangle
VULPECULA
PEGASUS
Rasalhague
OPHIUCHUS
Ring Nebula
LYRA
Sadr
CYGNUS
Jupiter
Deneb
Vega
Zenith
LACERTA
Sarin
Alpheratz
PISCES
HERCULES
Eltanin
Alderamin
ζ Her
M13
M92
Andromeda Galaxy
SERPENS
DRACO
Caph
Mirach
ANDROMEDA
Triangulum Galaxy
CETUS
CORONA BOREALIS
URSA MINOR
CASSIOPEIA
Almach
Hamal
Polaris
Double Cluster
TRIANGULUM
ARIES
Izar
Kochab
W
E
Algol
BOÖTES
Alkaid
Mizar
CAMELOPARDALIS
Mirfak
PERSEUS
Arcturus
Whirlpool Galaxy
Alioth
URSA MAJOR
Uranus
Dubhe
Pleiades
CANES VENATICI
Capella
Ecliptic
Merak
LYNX
AURIGA
NW
NE
N
CONSTELLATIONS
Galaxies
Star Clusters
Nebulæ
Planets
Stars
Region of interest
September 1: 11:00 p.m. (12:00 a.m. DST)
September 15: 10:00 p.m. (11:00 p.m. DST)
October 1: 9:00 p.m. (10:00 p.m. DST)

# Highlights in the Northern Sky

The **Big Dipper** now sits low in the northwest along with **Boötes (the Herdsman)** and **Corona Borealis (the Northern Crown)**.

**Hercules** sits above Corona Borealis, which means it's an opportune time to check out the globular clusters **Messier 13** and **Messier 92**.

You can still view the **Double Cluster (NGC 869 and NGC 884)**, as **Perseus** is beginning to rise in the northeast. If you're observing using a pair of binoculars, be sure to find **Mirfak**, the brightest star in Perseus, which holds a lovely collection of bright stars.

And, of course, it's always worth observing the **Andromeda Galaxy (Messier 31)**, whether it's with the unaided eye in dark-sky conditions or through binoculars or a telescope.

You can also turn your eyes to the zenith to find **Cygnus (the Swan)**. Check out its abundance of stars, which are fantastic when viewed with binoculars and make a lovely backdrop to the end of summer.

The globular cluster Messier 92

# October

## October's Events

The nights are longer and cooler at this time of year, but that also means there's less atmospheric interference in the sky, so the stars may seem brighter.

Early in the month, if you are a lark, you can find Mercury low near the eastern horizon just before sunrise. It will be fairly bright, but you can always use a pair of binoculars to enhance the view.

Jupiter shines brightly and high in the evening sky this month. Saturn is also visible, though not quite as high in the sky as Jupiter.

The summer constellations are now setting in the west, but that also means the beautiful winter constellations are starting to creep up in the sky.

And finally, this month we are treated to another meteor shower with the Orionids. The peak night is October 21 to 22. It's not one of the best showers of the year, but it can produce about 10 to 20 meteors an hour at its peak.

## Calendar of Events

| Day | Time (UTC) | Event |
|---|---|---|
| 3 | 00:14 | First quarter of the Moon |
| 4 | 17:01 | Moon at perigee: 369,300 km (229,472 mi.) |
| 5 | 15:51 | Saturn 4°N of Moon |
| 8 | 02:33 | Neptune 3.1°N of Moon |
| 8 | 18:06 | Jupiter 2°N of Moon |
| 8 | 21:13 | Mercury 18°W of Sun (greatest western elongation) |
| 9 | 20:55 | Full Moon |
| 12 | 06:46 | Uranus 0.9°S of Moon |
| 13 | 03:46 | Moon 2.9°S of Pleiades |
| 14 | 04:28 | Mars 3°S of Moon |
| 17 | 10:21 | Moon at apogee: 404,300 km (251,220 mi.) |
| 17 | 17:15 | Last quarter of the Moon |
| 21–22 | | Orionid meteor shower peak |
| 22 | 20:47 | Venus at superior conjunction |
| 24 | 15:03 | Mercury 0.2°S of Moon |
| 25 | 10:49 | New Moon |
| 25 | 11:00 | Partial solar eclipse (not observable from North America) |
| 29 | 14:48 | Moon at perigee: 368,300 km (228,851 mi.) |

## The Moon This Month

| SUN | MON | TUES | WED | THURS | FRI | SAT |
|---|---|---|---|---|---|---|
| | | | | | | 1 |
| 2 | 3 1st Quarter | 4 | 5 | 6 | 7 | 8 |
| 9 Full Moon | 10 | 11 | 12 | 13 | 14 | 15 |
| 16 | 17 Last Quarter | 18 | 19 | 20 | 21 | 22 |
| 23 | 24 | 25 New Moon | 26 | 27 | 28 | 29 |
| 30 | 31 | | | | | |

On the night of October 5, the Moon passes about 4 degrees south of Saturn. For a challenge, on the night of October 7, use binoculars to find the planet Neptune just above the waxing gibbous Moon. On October 8, the Moon will be roughly 3 degrees from our Solar System's giant, Jupiter.

On the night of October 11, use binoculars to locate Uranus very near the waning gibbous Moon. Some lucky observers will get to see the Moon pass in front of the planet, a phenomenon called an "occultation." Check with your local astronomy club for details regarding your chances of witnessing this rare event.

On October 14, Mars gets a close encounter when the Moon is about 2 degrees from the red planet. Both rise in the east around 10:00 p.m. local time. If you are up for an observing challenge, just before sunrise on October 24 use binoculars to look for a very old Moon – a thin waning crescent low in the east, with Mercury very close by. For most of North America, this rising Moon is less than 24 hours from being new, a popular observing goal. Some observers may even see an occultation of Mercury by the Moon. Check with your local astronomy club.

Perigee occurs on both October 4 and October 29, with apogee falling on October 17. The full Moon is on October 9, and the new Moon occurs on October 25.

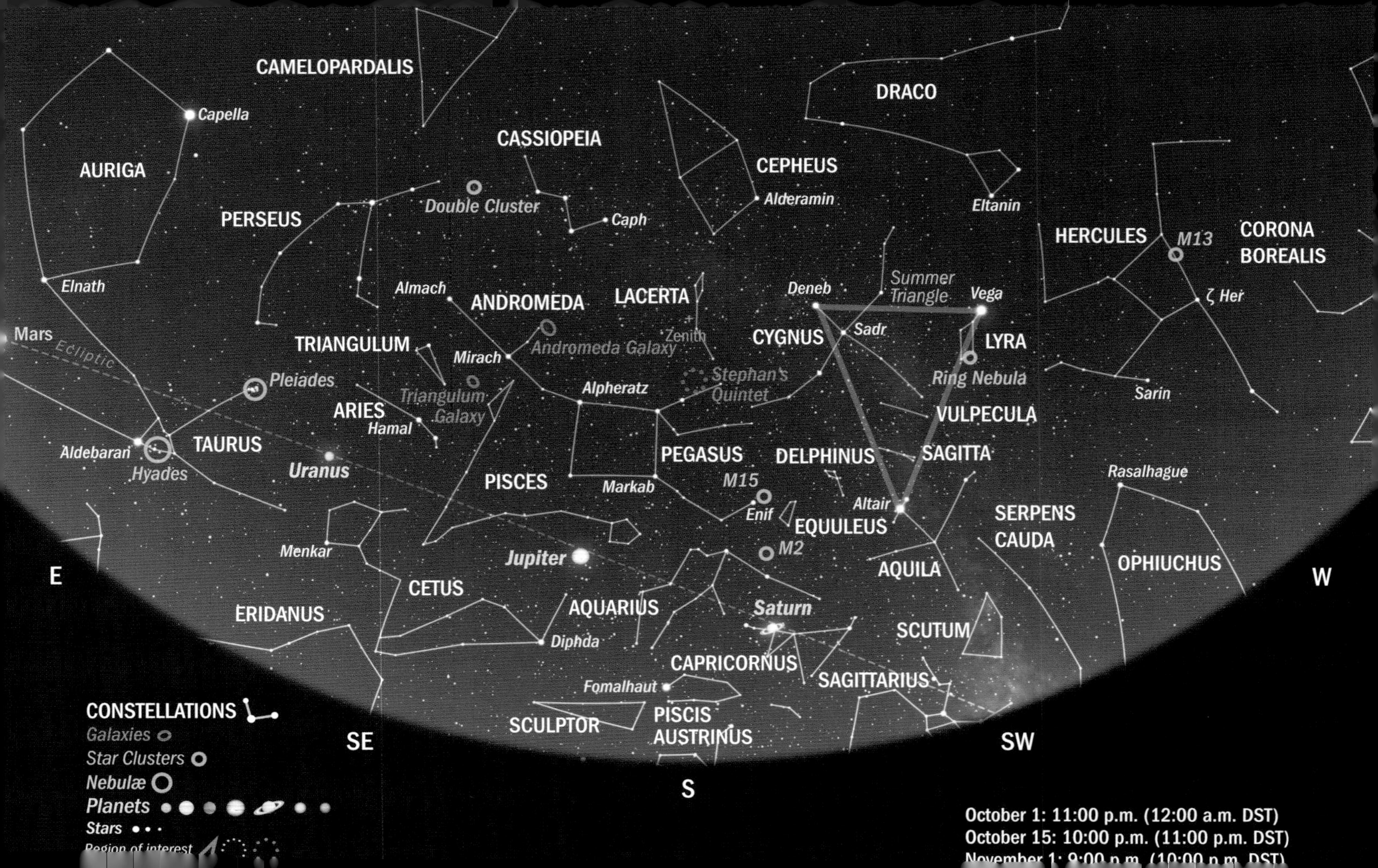
CAMELOPARDALIS
AURIGA
Capella
Elnath
PERSEUS
Double Cluster
CASSIOPEIA
Caph
CEPHEUS
Alderamin
DRACO
Eltanin
HERCULES
M13
CORONA BOREALIS
ζ Her
Sarin
Mars
Ecliptic
Pleiades
TRIANGULUM
Almach
ANDROMEDA
LACERTA
Zenith
Deneb
Summer Triangle
Vega
CYGNUS
Sadr
LYRA
Ring Nebula
Mirach
Andromeda Galaxy
Stephan's Quintet
Triangulum Galaxy
ARIES
Hamal
Alpheratz
VULPECULA
Aldebaran
Hyades
TAURUS
Uranus
PEGASUS
DELPHINUS
SAGITTA
Rasalhague
PISCES
Markab
M15
Enif
Altair
EQUULEUS
SERPENS CAUDA
Menkar
Jupiter
M2
AQUILA
OPHIUCHUS
E
W
CETUS
ERIDANUS
AQUARIUS
Saturn
SCUTUM
Diphda
CAPRICORNUS
SAGITTARIUS
Fomalhaut
SCULPTOR
PISCIS AUSTRINUS
SE
S
SW
CONSTELLATIONS
Galaxies
Star Clusters
Nebulæ
Planets
Stars
October 1: 11:00 p.m. (12:00 a.m. DST)
October 15: 10:00 p.m. (11:00 p.m. DST)

# Highlights in the Southern Sky

If you look to the south, you'll find **Saturn** sitting in **Capricornus (the Sea Goat)** and Jupiter between **Pisces (the Fishes)** and **Cetus (the Whale)**.

You will also notice a fairly bright star low on the southern horizon. That's **Fomalhaut**, the brightest star in **Piscis Austrinus (the Southern Fish)**. The star is the 17th brightest in the sky and is located just 25 light-years from Earth. While at one time it was believed that Hubble had captured an image of an exoplanet amid the dust surrounding the star, it had disappeared by 2020, leading astronomers to believe that what Hubble had photographed was an expanding cloud of dust left over from the collision of two icy bodies.

High in the southeast you'll notice a large square gracing the sky. That's the square of **Pegasus**. The constellation is the seventh largest in the sky and has some beautiful targets, including **Messier 15** and **Stephan's Quintet**, a collection of five galaxies. It's best to look at these using a telescope.

Between Jupiter and Saturn is **Aquarius (the Water Bearer)**. In this constellation you can find a beautiful globular cluster, **Messier 2**. It is roughly 37,000 light-years from Earth and has a diameter of approximately 150 light-years, making it one of the largest globular clusters in the sky.

**Deneb**, the brightest star in **Cygnus (the Swan)**, is almost directly overhead, which also means that the **Summer Triangle** is still well-placed for observing a rich part of the Milky Way. **Aquila (the Eagle)** is beginning to set in the west along with **Scutum (the Shield)**.

Stephan's Quintet

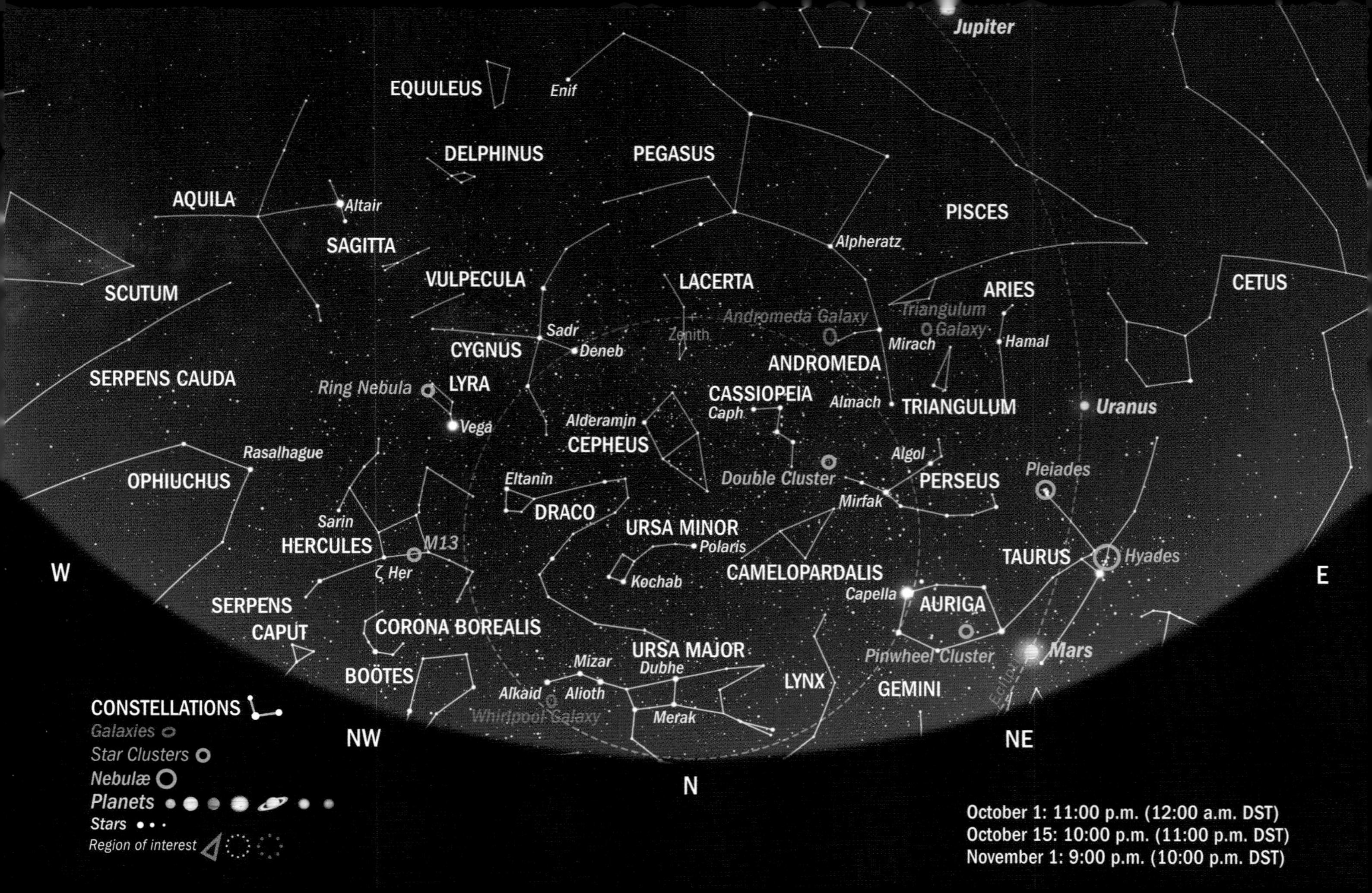
Jupiter
EQUULEUS
Enif
DELPHINUS
PEGASUS
PISCES
AQUILA
Altair
SAGITTA
Alpheratz
SCUTUM
VULPECULA
LACERTA
ARIES
CETUS
Andromeda Galaxy
Triangulum Galaxy
Sadr
Zenith
Deneb
Mirach
Hamal
CYGNUS
ANDROMEDA
SERPENS CAUDA
Ring Nebula
LYRA
CASSIOPEIA
Caph
Almach
TRIANGULUM
Uranus
Vega
Alderamin
CEPHEUS
Algol
Rasalhague
Double Cluster
Pleiades
OPHIUCHUS
Eltanin
PERSEUS
Mirfak
DRACO
Sarin
URSA MINOR
HERCULES
M13
Polaris
TAURUS
Hyades
W
ζ Her
Kochab
CAMELOPARDALIS
E
Capella
SERPENS
AURIGA
CAPUT
CORONA BOREALIS
URSA MAJOR
Mars
Mizar
Pinwheel Cluster
Dubhe
BOÖTES
LYNX
GEMINI
Alkaid
Alioth
Ecliptic
Whirlpool Galaxy
Merak
CONSTELLATIONS
Galaxies
NW
NE
Star Clusters
Nebulæ
N
Planets
Stars
Region of interest
October 1: 11:00 p.m. (12:00 a.m. DST)
October 15: 10:00 p.m. (11:00 p.m. DST)
November 1: 9:00 p.m. (10:00 p.m. DST)

# Highlights in the Northern Sky

**Ursa Major (the Great Bear)** is now low in the north, with its asterism the **Big Dipper** featuring prominently.

To the northeast is **Auriga (the Charioteer)**, with its star **Capella** shining brightly. There are several star clusters to observe, including the **Pinwheel Cluster (Messier 36)**. This cluster, which contains about 60 stars, lies about 4,100 light-years from Earth and is best viewed through binoculars.

About three hours after sunset, you might notice a bright red "star" rising below Auriga, between the horns of **Taurus (the Bull)**. That's not a star, but **Mars**, which will begin to rise higher in the sky and toward the south by the end of the month and into November.

**Perseus** and **Cassiopeia** are both now higher in the northwest, which puts them in a great spot for observing the **Andromeda Galaxy (Messier 31)**.

To the northwest, **Hercules** is now beginning to move toward the horizon, so try to check out the stunning globular cluster **Messier 13** one last time before it disappears at the end of the month.

**Lyra (the Lyre)**, with its stunning bright star **Vega**, is also unmistakable in the northwest, though it, too, is beginning to sink toward the horizon.

If you have a telescope, check out the **Ring Nebula (Messier 57)**. This is a planetary nebula, left over after a Sun-like star blew out its outer layer and became a white dwarf. When our Sun is at the end of its life, it will also die in this manner.

The globular cluster Messier 13

# November

## November's Events

November means colder weather, but it also marks the return of the winter constellations and what is likely the most well-known constellation: Orion.

The winter sky has a lot to offer if you're willing to brave the chillier nights. There's the multitude of objects in Orion as well as those in nearby Taurus and Auriga. The cooler nights also mean that the sky seems less turbulent, allowing the colors of the many different stars to pop.

There are two meteor shower peaks this month, the Southern Taurids and the Northern Taurids. The Southern Taurid meteor shower peaks on November 5, with the Northern Taurids peaking a few days later, on November 11. With the two peaks occurring so close together, there's a greater chance to see meteors that week.

But that's not all when it comes to meteors. On the night of November 17 to 18, we get the peak of the Leonid meteor shower, which produces a ZHR of roughly 15, so this is a particularly good month to bundle up and look skyward.

## Calendar of Events

| Day | Time (UTC) | Event |
|---|---|---|
| 1 | 06:37 | First quarter of the Moon |
| 1 | 21:08 | Saturn 4°S of Moon |
| 4 | 20:19 | Jupiter 2°N of Moon |
| 5–6 | 18:08 | Southern Taurid meteor shower peak |
| 6 | | Daylight Saving Time ends (most of U.S. and Canada) |
| 8 | 10:59 | Total lunar eclipse |
| 8 | 11:02 | Full Moon |
| 8 | 16:28 | Mercury at superior conjunction |
| 11 | 13:43 | Mars 2°S of Moon |
| 11–12 | | Northern Taurid meteor shower peak |
| 14 | 06:41 | Moon at apogee: 404,900 km (251,593 mi.) |
| 16 | 13:27 | Last quarter of the Moon |
| 17–18 | | Leonid meteor shower peak |
| 23 | 22:57 | New Moon |
| 26 | 01:30 | Moon at perigee: 362,800 km (225,433 mi.) |
| 29 | 04:40 | Saturn 4°N of Moon |
| 30 | 14:36 | First quarter of the Moon |

## The Moon This Month

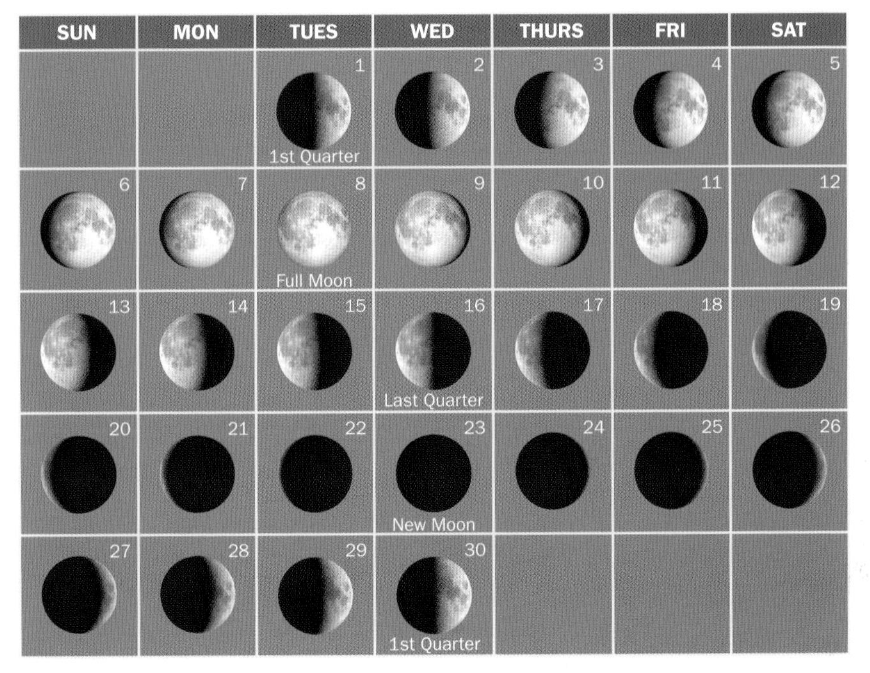


On November 1, the Moon will be roughly 4 degrees from Saturn in the southern sky, and on November 4 it will be about 2 degrees from Jupiter.

The full Moon will swing by the Pleiades and the Hyades on November 8 and then again on November 9. And, of course, the Moon can't miss out on visiting Mars: On November 11 the pair will be roughly 3 degrees apart.

There will be a total lunar eclipse in the early hours of November 8, which will be visible either in part or in its entirety across North America. (See page 28.)

The new Moon occurs on November 23. Apogee falls on November 14, and perigee falls on November 26.

The 2023 edition of *Night Sky Almanac* is in stores this month. Be sure to grab your copy, wherever books are sold.

A total lunar eclipse occurs this month

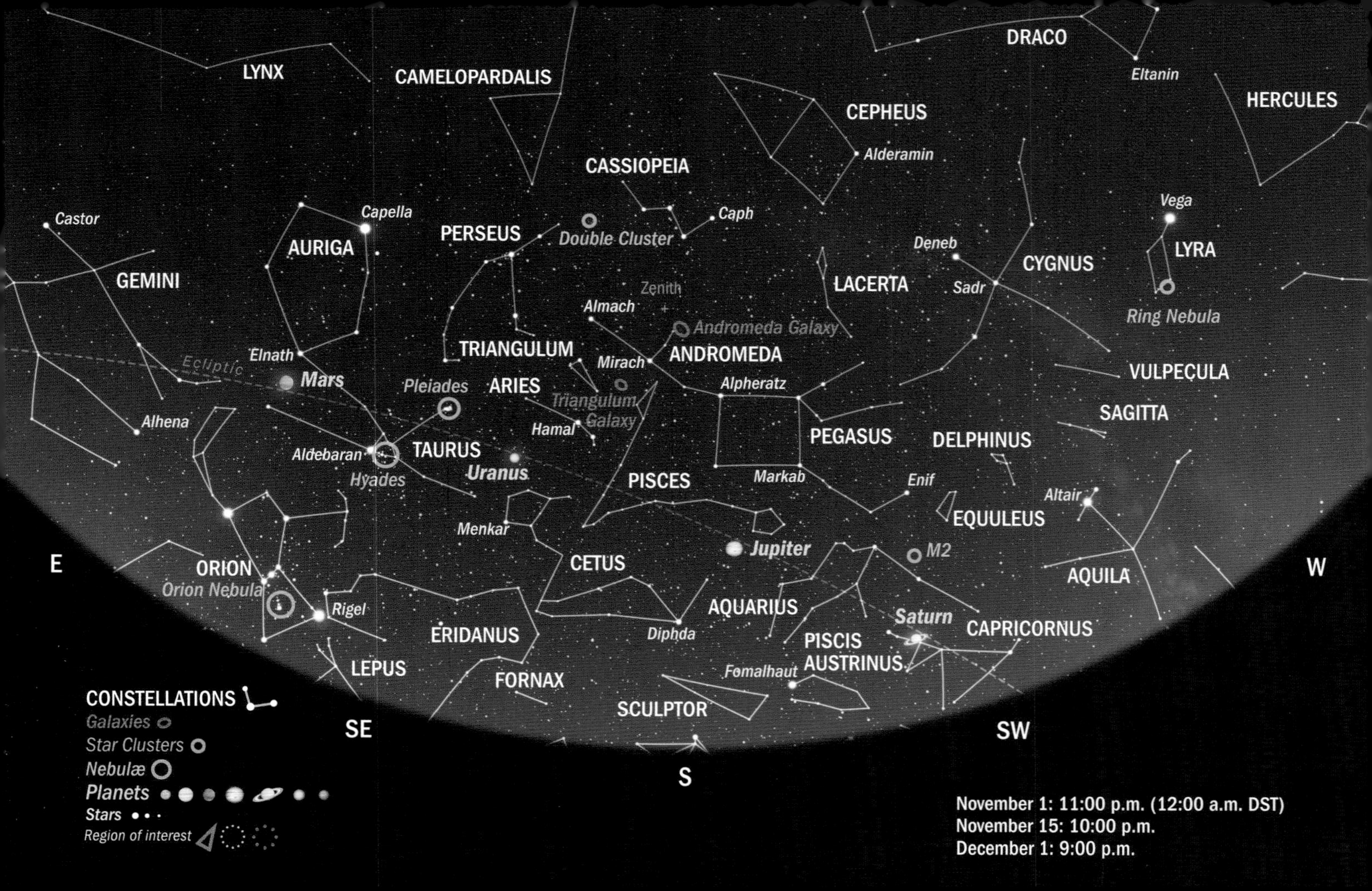

LYNX
CAMELOPARDALIS
DRACO
Eltanin
HERCULES
CEPHEUS
Alderamin
CASSIOPEIA
Caph
Double Cluster
Vega
LYRA
Ring Nebula
Deneb
Sadr
CYGNUS
LACERTA
Zenith
Almach
Andromeda Galaxy
ANDROMEDA
Mirach
Castor
GEMINI
AURIGA
Capella
PERSEUS
TRIANGULUM
Triangulum Galaxy
Elnath
Ecliptic
Mars
Pleiades
ARIES
Hamal
Alpheratz
PEGASUS
VULPECULA
SAGITTA
DELPHINUS
Alhena
Aldebaran
Hyades
TAURUS
Uranus
PISCES
Markab
Enif
Altair
EQUULEUS
Menkar
CETUS
Jupiter
M2
AQUILA
E
ORION
Orion Nebula
Rigel
ERIDANUS
AQUARIUS
Diphda
Saturn
CAPRICORNUS
PISCIS AUSTRINUS
W
LEPUS
FORNAX
Fomalhaut
SCULPTOR
SE
S
SW
CONSTELLATIONS
Galaxies
Star Clusters
Nebulæ
Planets
Stars
Region of interest
November 1: 11:00 p.m. (12:00 a.m. DST)
November 15: 10:00 p.m.
December 1: 9:00 p.m.

## Highlights in the Southern Sky

This month, you will find **Jupiter** sitting between **Cetus (the Whale)** and **Pisces (the Fishes)**. Saturn is in **Capricornus (the Sea Goat)**, though it begins to set earlier as the month progresses. **Mars** is beginning to rise higher in the sky, between the horns of **Taurus (the Bull)**. If you are wondering what happened to Venus and Mercury, they are far too close to the Sun to be seen.

One of the smaller but fairly obvious constellations is **Triangulum (the Triangle)**, which can be found amid the constellations **Perseus**, **Andromeda**, Pisces and **Aries (the Ram)**. Triangulum is home to one of the most famous galaxies, the **Triangulum Galaxy (Messier 33)**. Interestingly, the galaxy is believed to have interacted with the **Andromeda Galaxy (Messier 31)** – which can be found nearby in the constellation **Andromeda** – billions of years ago and is now moving toward it.

In the west, you will find **Aquarius (the Water Bearer)**, where you can see the globular cluster **Messier 2** before it sets at month's end.

This is a great time to check out two of the most famous open star clusters of the winter sky, the **Hyades (Melotte 25)** and the **Pleiades (Messier 45)**, which are found in Taurus. The reddish star **Aldebaran** is the brightest star in Taurus and is easily visible even in light-polluted skies above cities.

The famous **Orion (the Hunter)** will rise in the east as the night progresses, along with **Gemini (the Twins)**.

The Triangulum Galaxy (Messier 33)

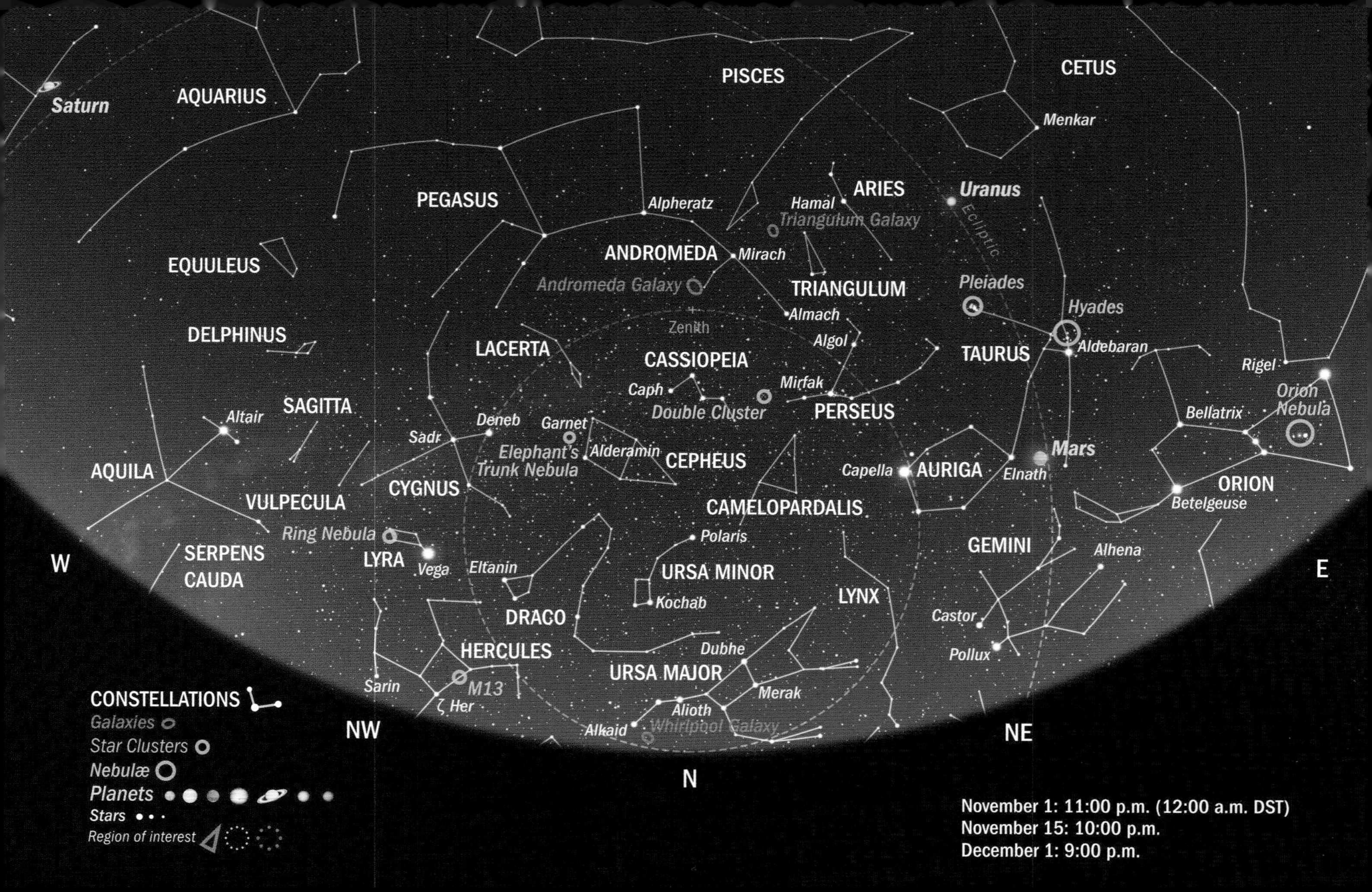

Saturn
AQUARIUS
PISCES
CETUS
Menkar
PEGASUS
Alpheratz
ARIES
Hamal
Uranus
Triangulum Galaxy
Ecliptic
EQUULEUS
ANDROMEDA
Mirach
Andromeda Galaxy
TRIANGULUM
Pleiades
Hyades
Almach
Zenith
DELPHINUS
LACERTA
CASSIOPEIA
Algol
TAURUS
Aldebaran
Caph
Mirfak
Rigel
Orion Nebula
SAGITTA
Altair
Double Cluster
PERSEUS
Bellatrix
Deneb
Garnet
Sadr
Elephant's Trunk Nebula
Alderamin
CEPHEUS
Mars
AQUILA
Capella
AURIGA
Elnath
ORION
CYGNUS
VULPECULA
CAMELOPARDALIS
Betelgeuse
Ring Nebula
Polaris
SERPENS CAUDA
LYRA
Vega
Eltanin
URSA MINOR
GEMINI
Alhena
W
E
Kochab
LYNX
DRACO
Castor
HERCULES
Dubhe
Pollux
URSA MAJOR
Sarin
M13
Merak
ζ Her
Alioth
Alkaid
Whirlpool Galaxy
NW
NE
N
CONSTELLATIONS
Galaxies
Star Clusters
Nebulæ
Planets
Stars
Region of interest
November 1: 11:00 p.m. (12:00 a.m. DST)
November 15: 10:00 p.m.
December 1: 9:00 p.m.

## Highlights in the Northern Sky

The bowl of the **Big Dipper** in **Ursa Major (the Great Bear)** is now very low on the horizon in the north, and **Ursa Minor (the Little Bear)** hangs upside down due north.

**Draco (the Dragon)** winds between the two bears and leads to **Hercules**, which is now beginning to sink lower in the northwest.

**Cassiopeia**'s distinctive W hangs almost upside down between **Cepheus** and **Perseus**. This is another good opportunity to enjoy the **Garnet Star** in Cepheus (see page 89). The constellation is also home to the **Elephant's Trunk Nebula (IC 1396)**, which is a dense region of dust and gas that is busy forming new stars. It's located roughly 2,400 light-years from Earth. It is also a favorite target of astrophotographers.

And if you're looking for the **Andromeda Galaxy (Messier 31)**, it's almost right at the zenith. The placement of Cassiopeia, Cepheus and Perseus provide some excellent opportunities to find not only the Andromeda Galaxy, but star clusters as well, including the **Double Cluster (NGC 869 and NGC 884)**. Be sure to take out your binoculars for a great view.

**Auriga** has now risen higher in the northeast along with the two heads of **Gemini (the Twins)**, **Castor** and **Pollux**.

The Double Cluster (NGC 869 and NGC 884)

# December

## December's Events

This month, three planets will make their way across the sky: Mars, Jupiter and Saturn. Mars rises later in the night, while Saturn is setting. Meanwhile, Jupiter remains high in the sky for optimal viewing. Mercury and Venus are now together, low in the southwest after sunset, but they will be easier to find later in the month, and you may need binoculars to spot Mercury in the twilight.

December also means it's time for the Geminid meteor shower, which peaks on the night of December 13 to 14. This is the most active shower of the year, though seeing it can be a gamble since it tends to be cloudier during December. The peak can produce upward of 150 meteors an hour. There's also a second shower that tends to get overlooked, the Ursids, which peaks on the night of December 22 to 23. This shower produces about 10 meteors an hour.

Of course, aside from all the planetary and meteoric activity, there are still plenty of nebulae, stars and clusters to enjoy.

## Calendar of Events

| Day | Time (UTC) | Event |
|---|---|---|
| 2 | 00:52 | Jupiter 3°N of Moon |
| 8 | 04:08 | Full Moon |
| 8 | 04:21 | Mars 0.5°S of Moon |
| 8 | 04:24 | Mars at opposition |
| 12 | 00:30 | Moon at apogee: 405,900 km (252,215 mi.) |
| 13–14 | | Geminid meteor shower peak |
| 16 | 08:56 | Last quarter of the Moon |
| 21 | 15:31 | Mercury 20.1°E of Sun (greatest eastern elongation) |
| 21 | 21:48 | Winter solstice |
| 22–23 | | Ursid meteor shower peak |
| 23 | 10:17 | New Moon |
| 24 | 08:32 | Moon at perigee: 358,300 km (222,637 mi.) |
| 24 | 11:29 | Venus 3°N of Moon |
| 24 | 18:30 | Mercury 4°N of Moon |
| 26 | 16:11 | Saturn 4°N of Moon |
| 28 | 20:02 | Neptune 3°N of Moon |
| 29 | 07:13 | Mercury 1.4°N of Venus |
| 29 | 10:29 | Jupiter 2°N of Moon |
| 30 | 01:20 | First quarter of the Moon |

## The Moon This Month

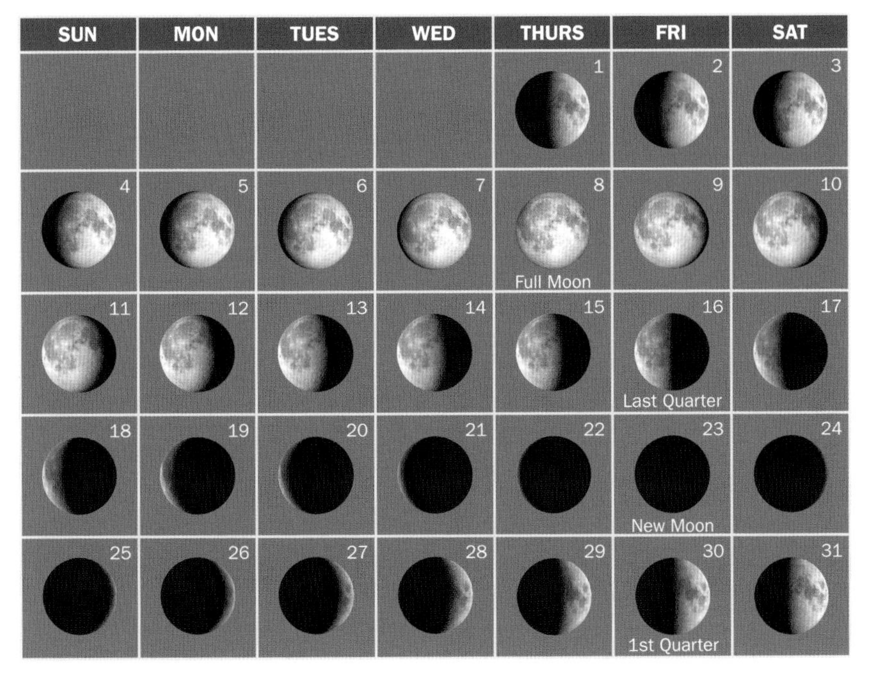


On December 2, a waxing gibbous Moon joins Jupiter in the southern sky, where they will be about 3 degrees apart. The Moon meets the Pleiades on December 6.

On the evening of December 7, most of North America can witness the full Moon pass in front of Mars. This rare lunar occultation event is made even rarer because Mars is also at opposition; however, the opposition of Mars will not be as spectacular as the previous one in October 2020. The orbit of Mars is eccentric, and this year's "close approach" will be 32 percent more distant, with the disk correspondingly smaller. It will appear only 17 seconds of arc across, less than half the apparent width of Jupiter. Mars will also not be as bright as Jupiter. Even so, in a telescope it will be exciting to watch the Moon cross the disk of Mars and then watch Mars reappear from behind the Moon as much as an hour later. Check with your local astronomy club for more details.

For a challenge and a treat, find a low southwestern horizon just after sunset on December 24 and, using binoculars, look for the trio of Venus, Mercury and a two-day old crescent Moon.

On December 26, the Moon joins Saturn as a crescent just after sunset in the southwest, where they will be roughly 4 degrees apart. Two nights later, on the evening of December 28, you have another chance (using binoculars) to find the planet Neptune above and to the right of the Moon. A quarter Moon swings past Jupiter again at the end of the month, on December 29. They will be about 5 degrees apart.

The full Moon falls on December 8 and the new Moon on December 23. Apogee occurs on December 12 and perigee on December 24.

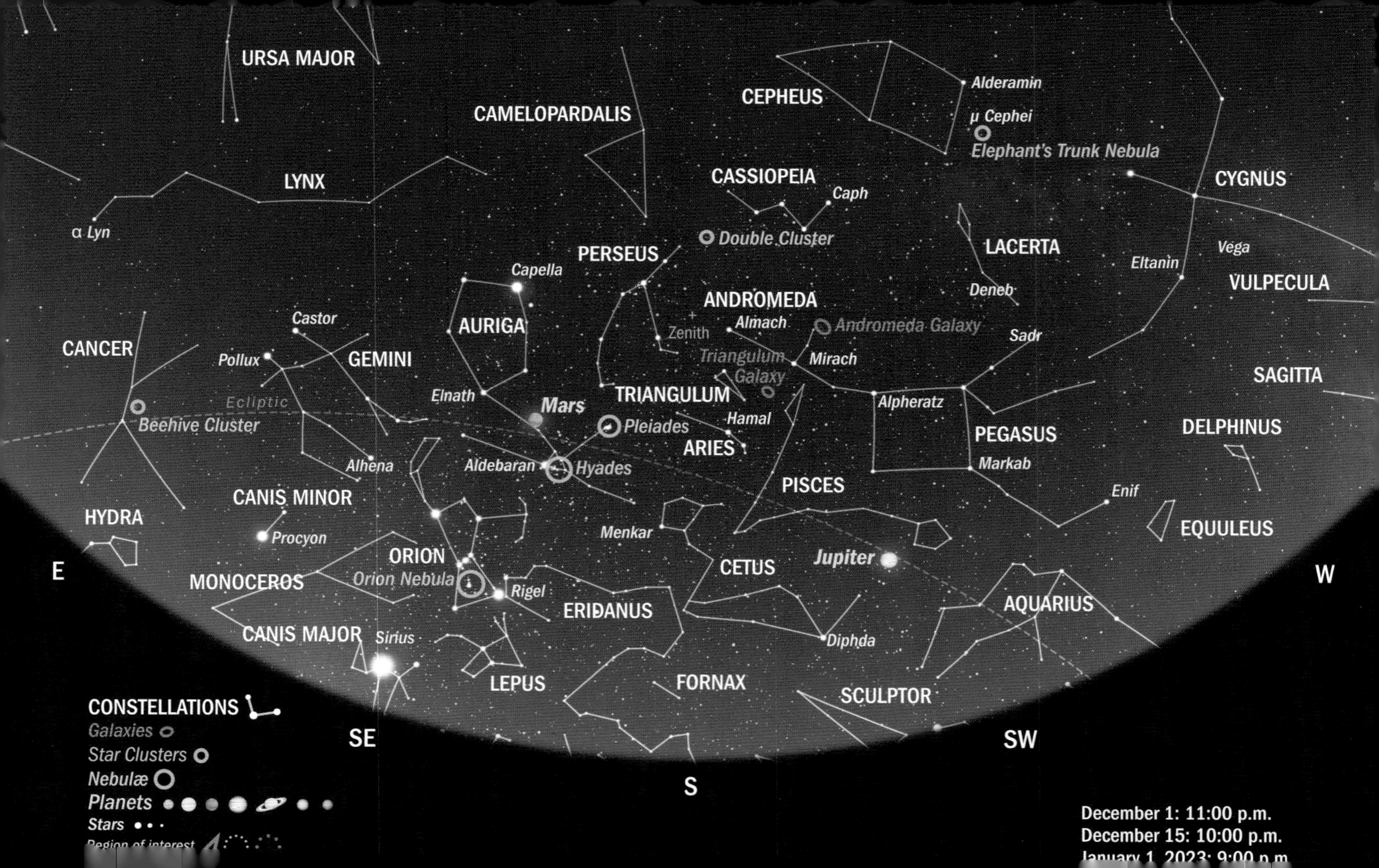
URSA MAJOR
CAMELOPARDALIS
CEPHEUS
Alderamin
μ Cephei
Elephant's Trunk Nebula
LYNX
CASSIOPEIA
Caph
CYGNUS
α Lyn
Double Cluster
PERSEUS
LACERTA
Vega
Eltanin
Capella
Deneb
VULPECULA
ANDROMEDA
Castor
AURIGA
Zenith
Almach
Andromeda Galaxy
Sadr
CANCER
Pollux
GEMINI
Triangulum Galaxy
Mirach
SAGITTA
Elnath
Mars
TRIANGULUM
Alpheratz
Ecliptic
Beehive Cluster
Pleiades
Hamal
PEGASUS
DELPHINUS
ARIES
Alhena
Aldebaran
Hyades
Markab
PISCES
CANIS MINOR
Enif
HYDRA
Menkar
EQUULEUS
Procyon
ORION
Jupiter
E
CETUS
W
MONOCEROS
Orion Nebula
Rigel
AQUARIUS
ERIDANUS
CANIS MAJOR
Sirius
Diphda
LEPUS
FORNAX
SCULPTOR
CONSTELLATIONS
Galaxies
Star Clusters
Nebulæ
Planets
Stars
Region of interest
SE
SW
S
December 1: 11:00 p.m.
December 15: 10:00 p.m.
January 1, 2023: 9:00 p.m.

# Highlights in the Southern Sky

**Orion (the Hunter)** is front and center this month in the southern sky. There's a lot to see, even with the naked eye, including the **Orion Nebula (Messier 42)**, a beautiful star-forming region just below the stars of the hunter's belt.

It's also fun to check out various stars, like **Betelgeuse**, which represents Orion's left shoulder, and **Rigel**, which represents Orion's right foot.

Then there's the reddish star **Aldebaran** in **Taurus (the Bull)**. However, this month you might notice another, brighter red "star." That would be **Mars**, which sits in Taurus. The planet will be at its brightest this month, beginning to dim somewhat by the end of the month. Don't miss the opposition of Mars on the evening of December 7, with the possibility of a lunar occultation of the planet.

You can also grab a pair of binoculars and find the open star clusters the **Hyades (Melotte 25)** and the **Pleiades (Messier 45)**.

**Eridanus (the River Eridanus)**, one of the longest constellations, is now above the horizon. Beginning near Rigel in Orion, the constellation represented a river in many cultures. For ancient Egyptians, it was the Nile or the Euphrates; for Romans, the Po; and in ancient China, the Huang He (also known as the Yellow River).

**Sirius**, the brightest star in our night sky, is part of the constellation **Canis Major (the Great Dog)**. It shines brightly along the southern horizon with its dimmer cousin **Procyon**, which is part of the constellation **Canis Minor (the Little Dog)**.

**Cancer (the Crab)** can be found in the east along with the **Beehive Cluster (Messier 44)** – a great binocular target.

The star Sirius

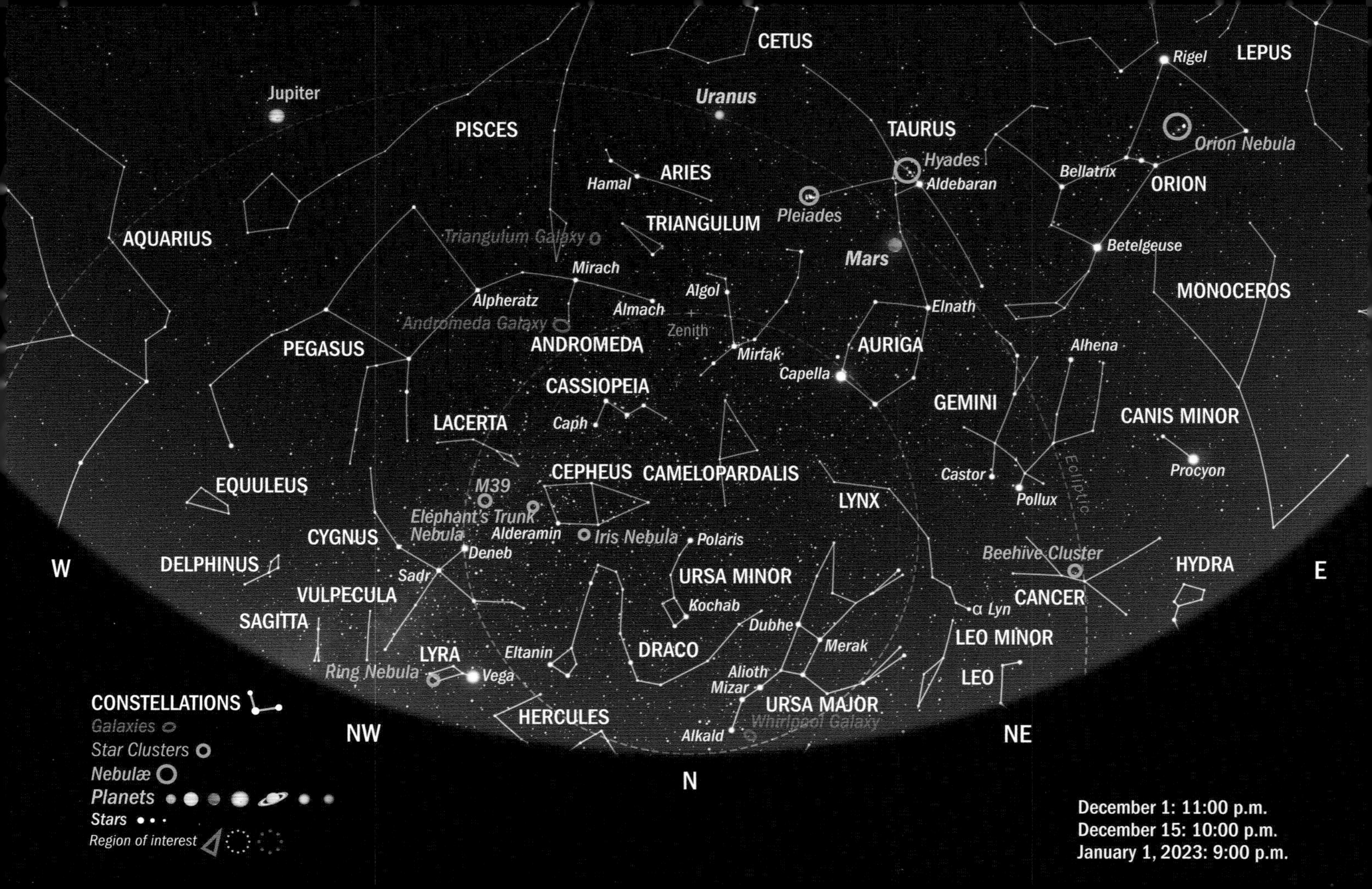
CETUS
LEPUS
Jupiter
Uranus
PISCES
TAURUS
Rigel
Orion Nebula
Hyades
Aldebaran
Bellatrix
ORION
ARIES
Hamal
AQUARIUS
TRIANGULUM
Pleiades
Triangulum Galaxy
Mars
Betelgeuse
Mirach
Algol
MONOCEROS
Alpheratz
Almach
Elnath
Andromeda Galaxy
Zenith
PEGASUS
ANDROMEDA
AURIGA
Mirfak
Alhena
Capella
CASSIOPEIA
GEMINI
CANIS MINOR
LACERTA
Caph
Castor
Procyon
EQUULEUS
CEPHEUS
CAMELOPARDALIS
M39
Pollux
LYNX
Ecliptic
Elephant's Trunk Nebula
Alderamin
Iris Nebula
Polaris
CYGNUS
Deneb
Beehive Cluster
W
DELPHINUS
HYDRA
E
Sadr
URSA MINOR
VULPECULA
α Lyn
CANCER
Kochab
SAGITTA
Dubhe
LEO MINOR
Merak
LYRA
Eltanin
DRACO
Alioth
Ring Nebula
Vega
LEO
Mizar
URSA MAJOR
CONSTELLATIONS
HERCULES
Whirlpool Galaxy
Galaxies
Alkald
NW
NE
Star Clusters
N
Nebulæ
Planets
Stars
Region of interest
December 1: 11:00 p.m.
December 15: 10:00 p.m.
January 1, 2023: 9:00 p.m.

## Highlights in the Northern Sky

At this time of year, the **Big Dipper** asterism in **Ursa Major (the Great Bear)** is now beginning its upward journey in the sky.

The outstretched wings of **Cygnus (the Swan)** reach out in the west. Its brightest star, **Deneb**, which marks the tail of the swan, shines brightly. The constellation is sometimes referred to as the Northern Cross, similar to the Southern Cross in the Southern Hemisphere. (The latter constellation is featured on several flags, including those of Australia and New Zealand.)

You might want to grab a pair of binoculars to take one last look at the rich region in Cygnus before it dips below the horizon, and be sure to look for **Messier 39**, an open star cluster not far from Deneb.

**Cassiopeia** is now hanging upside down high above **Cepheus**, the little house lying on its side.

**Camelopardalis (the Giraffe)** can be found near Cassiopeia and **Perseus**, and the star **Capella** is unmistakable in the constellation **Auriga (the Charioteer)**.

The Big Dipper asterism acts as the background to Comet NEOWISE C/2020 F3

# The Messier Catalog

| Name | Traditional Name | Type | Constellation |
| --- | --- | --- | --- |
| Messier 1 (NGC 1952) | Crab Nebula | Supernova remnant | Taurus |
| Messier 2 (NGC 7089) | | Globular cluster | Aquarius |
| Messier 3 (NGC 5272) | | Globular cluster | Canes Venatici |
| Messier 4 (NGC 6121) | | Globular cluster | Scorpius |
| Messier 5 (NGC 5904) | | Globular cluster | Serpens |
| Messier 6 (NGC 6405) | Butterfly Cluster | Open cluster | Scorpius |
| Messier 7 (NGC 6475) | Ptolemy Cluster | Open cluster | Scorpius |
| Messier 8 (NGC 6523) | Lagoon Nebula | Emission nebula with cluster | Sagittarius |
| Messier 9 (NGC 6333) | | Globular cluster | Ophiuchus |
| Messier 10 (NGC 6254) | | Globular cluster | Ophiuchus |
| Messier 11 (NGC 6705) | Wild Duck Cluster | Open cluster | Scutum |
| Messier 12 (NGC 6218) | Gumball Globular | Globular cluster | Ophiuchus |
| Messier 13 (NGC 6205) | Hercules | Globular cluster | Hercules |
| Messier 14 (NGC 6402) | | Globular cluster | Ophiuchus |
| Messier 15 (NGC 7078) | Great Pegasus Cluster | Globular cluster | Pegasus |
| Messier 16 (NGC 6611) | Eagle Nebula | Emission nebula with cluster | Serpens |
| Messier 17 (NGC 6618) | Omega Nebula | Emission nebula with cluster | Sagittarius |
| Messier 18 (NGC 6613) | | Open cluster | Sagittarius |
| Messier 19 (NGC 6273) | | Globular cluster | Ophiuchus |
| Messier 20 (NGC 6514) | Trifid Nebula | Emission, reflection and dark nebula with cluster | Sagittarius |
| Messier 21 (NGC 6531) | | Open cluster | Sagittarius |
| Messier 22 (NGC 6656) | Sagittarius Cluster | Globular cluster | Sagittarius |
| Messier 23 (NGC 6494) | | Open cluster | Sagittarius |
| Messier 24 (IC 4715) | Sagittarius Star Cloud | Milky Way star cloud | Sagittarius |
| Messier 25 (IC 4725) | | Open cluster | Sagittarius |
| Messier 26 (NGC 6694) | | Open cluster | Scutum |
| Messier 27 (NGC 6853) | Dumbbell Nebula | Planetary nebula | Vulpecula |
| Messier 28 (NGC 6626) | | Globular cluster | Sagittarius |
| Messier 29 (NGC 6913) | | Open cluster | Cygnus |

| Name | Traditional Name | Type | Constellation |
| --- | --- | --- | --- |
| Messier 30 (NGC 7099) | | Globular cluster | Capricornus |
| Messier 31 (NGC 224) | Andromeda Galaxy | Spiral galaxy | Andromeda |
| Messier 32 (NGC 221) | Le Gentil | Dwarf elliptical galaxy | Andromeda |
| Messier 33 (NGC 598) | Triangulum Galaxy | Spiral galaxy | Triangulum |
| Messier 34 (NGC 1039) | | Open cluster | Perseus |
| Messier 35 (NGC 2168) | | Open cluster | Gemini |
| Messier 36 (NGC 1960) | Pinwheel Cluster | Open cluster | Auriga |
| Messier 37 (NGC 2099) | | Open cluster | Auriga |
| Messier 38 (NGC 1912) | Starfish Cluster | Open cluster | Auriga |
| Messier 39 (NGC 7092) | | Open cluster | Cygnus |
| Messier 40 | Winnecke 4 | Double star | Ursa Major |
| Messier 41 (NGC 2287) | | Open cluster | Canis Major |
| Messier 42 (NGC 1976) | Orion Nebula | Emission-reflection nebula | Orion |
| Messier 43 (NGC 1982) | De Mairan's Nebula | Emission-reflection nebula | Orion |
| Messier 44 (NGC 2632) | Beehive Cluster | Open cluster | Cancer |
| Messier 45 | Pleiades | Open cluster | Taurus |
| Messier 46 (NGC 2437) | | Open cluster | Puppis |
| Messier 47 (NGC 2422) | | Open cluster | Puppis |
| Messier 48 (NGC 2548) | | Open cluster | Hydra |
| Messier 49 (NGC 4472) | | Elliptical galaxy | Virgo |
| Messier 50 (NGC 2323) | Heart-Shaped Cluster | Open cluster | Monoceros |
| Messier 51 (NGC 5194, NGC 5195) | Whirlpool Galaxy | Spiral galaxy | Canes Venatici |
| Messier 52 (NGC 7654) | | Open cluster | Cassiopeia |
| Messier 53 (NGC 5024) | | Globular cluster | Coma Berenices |
| Messier 54 (NGC 6715) | | Globular cluster | Sagittarius |
| Messier 55 (NGC 6809) | Summer Rose Star | Globular cluster | Sagittarius |
| Messier 56 (NGC 6779) | | Globular cluster | Lyra |
| Messier 57 (NGC 6720) | Ring Nebula | Planetary nebula | Lyra |
| Messier 58 (NGC 4579) | | Barred spiral galaxy | Virgo |
| Messier 59 (NGC 4621) | | Elliptical galaxy | Virgo |

| Name | Traditional Name | Type | Constellation |
|---|---|---|---|
| Messier 60 (NGC 4649) | | Elliptical galaxy | Virgo |
| Messier 61 (NGC 4303) | | Spiral galaxy | Virgo |
| Messier 62 (NGC 6266) | | Globular cluster | Ophiuchus |
| Messier 63 (NGC 5055) | Sunflower Galaxy | Spiral galaxy | Canes Venatici |
| Messier 64 (NGC 4826) | Black Eye Galaxy | Spiral galaxy | Coma Berenices |
| Messier 65 (NGC 3623) | | Barred spiral galaxy | Leo |
| Messier 66 (NGC 3627) | | Barred spiral galaxy | Leo |
| Messier 67 (NGC 2682) | King Cobra Cluster | Open cluster | Cancer |
| Messier 68 (NGC 4590) | | Globular cluster | Hydra |
| Messier 69 (NGC 6637) | | Globular cluster | Sagittarius |
| Messier 70 (NGC 6681) | | Globular cluster | Sagittarius |
| Messier 71 (NGC 6838) | | Globular cluster | Sagittarius |
| Messier 72 (NGC 6981) | | Globular cluster | Aquarius |
| Messier 73 (NGC 6994) | | Asterism | Aquarius |
| Messier 74 (NGC 628) | Phantom Galaxy | Spiral galaxy | Pisces |
| Messier 75 (NGC 6864) | | Globular cluster | Sagittarius |
| Messier 76 (NGC 650, NGC 651) | Little Dumbbell Nebula | Planetary nebula | Perseus |
| Messier 77 (NGC 1068) | Cetus A | Spiral galaxy | Cetus |
| Messier 78 (NGC 2068) | | Reflection nebula | Orion |
| Messier 79 (NGC 1904) | | Globular cluster | Lepus |
| Messier 80 (NGC 6093) | | Globular cluster | Scorpius |
| Messier 81 (NGC 3031) | Bode's Galaxy | Spiral galaxy | Ursa Major |
| Messier 82 (NGC 3034) | Cigar Galaxy | Starburst irregular galaxy | Ursa Major |
| Messier 83 (NGC 5236) | Southern Pinwheel Galaxy | Barred spiral galaxy | Hydra |
| Messier 84 (NGC 4374) | | Lenticular or elliptical galaxy | Virgo |
| Messier 85 (NGC 4382) | | Lenticular or elliptical galaxy | Coma Berenices |
| Messier 86 (NGC 4406) | | Lenticular or elliptical galaxy | Virgo |
| Messier 87 (NGC 4486) | Virgo A | Elliptical galaxy | Virgo |

| Name | Traditional Name | Type | Constellation |
| --- | --- | --- | --- |
| Messier 88 (NGC 4501) | | Spiral galaxy | Coma Berenices |
| Messier 89 (NGC 4552) | | Elliptical galaxy | Virgo |
| Messier 90 (NGC 4569) | | Spiral galaxy | Virgo |
| Messier 91 (NGC 4548) | | Barred spiral galaxy | Coma Berenices |
| Messier 92 (NGC 6341) | | Globular cluster | Hercules |
| Messier 93 (NGC 2447) | | Open cluster | Puppis |
| Messier 94 (NGC 4736) | Cat's Eye Galaxy | Spiral galaxy | Canes Venatici |
| Messier 95 (NGC 3351) | | Barred spiral galaxy | Leo |
| Messier 96 (NGC 3368) | | Spiral galaxy | Leo |
| Messier 97 (NGC 3587) | Owl Nebula | Planetary nebula | Ursa Major |
| Messier 98 (NGC 4192) | | Spiral galaxy | Coma Berenices |
| Messier 99 (NGC 4254) | Coma Pinwheel | Spiral galaxy | Coma Berenices |
| Messier 100 (NGC 4321) | | Spiral galaxy | Coma Berenices |
| Messier 101 (NGC 5457) | Pinwheel Galaxy | Spiral galaxy | Ursa Major |
| Messier 102 (NGC 5866) | Spindle Galaxy | Lenticular galaxy | Draco |
| Messier 103 (NGC 581) | | Open cluster | Cassiopeia |
| Messier 104 (NGC 4594) | Sombrero Galaxy | Spiral galaxy | Virgo |
| Messier 105 (NGC 3379) | | Elliptical galaxy | Leo |
| Messier 106 (NGC 4258) | | Spiral galaxy | Canes Venatici |
| Messier 107 (NGC 6171) | | Globular cluster | Ophiuchus |
| Messier 108 (NGC 3556) | Surfboard Galaxy | Barred spiral galaxy | Ursa Major |
| Messier 109 (NGC 3992) | | Barred spiral galaxy | Ursa Major |
| Messier 110 (NGC 205) | Edward Young Star | Dwarf elliptical galaxy | Andromeda |

# Glossary

**annular solar eclipse:** A kind of partial solar eclipse during which the Moon does not completely cover the Sun's disk but leaves a ring of sunlight around the Moon.

**aphelion:** For an object orbiting the Sun, the point in its orbit that is farthest from the Sun.

**apogee:** When the Moon (or any satellite of Earth) is at its most distant in its monthly orbit around Earth.

**arcsecond (or second of arc):** A tiny angle equal to 1/3600 of a degree, used to measure the separation of double stars and the apparent diameters of solar system objects.

**asterism:** a group of stars within a constellation (or sometimes from more than one constellation) that forms its own distinct pattern.

**astronomical unit (AU):** A unit of distance that uses the average distance between Earth and the Sun as its base metric. One astronomical unit is approximately 150 million kilometers (93.2 million miles).

**autumnal (fall) equinox:** The date in September marking the end of summer and the beginning of autumn, when the Sun illuminates the Northern and Southern Hemispheres equally. The lengths of the day and night are equal, and it's one of two instances of the year that the Sun has a declination of 0 degrees.

**binary star system:** A double-star system in which two stars are gravitationally bound to each other and orbit a common center of mass. Many double-star systems have multiple components.

**conjunction:** Technically speaking a conjunction is when two objects have the same right ascension, but amateur astronomers also use the term when there is a close approach of two or more celestial objects.

**constellation:** A group of stars that make an imaginary image in the night sky.

**degree:** A unit of measurement used to measure the distance between objects or the position of objects in astronomy. The entire sky spans 360 degrees, and up to about 180 degrees of sky is visible from any given point on Earth with an unobstructed horizon.

**double star (binary star):** A pair of stars with small angular separation (from fractions of an arcsecond to hundreds of arcseconds). Visual doubles are chance alignments of stars with no physical connection to each other. True binary stars are gravitationally bound (see binary star system).

**fireball:** A very bright meteor, generally brighter than magnitude -4.0 (about the brightness of Venus).

**galaxy:** An enormous system of gas, dust and billions of stars and their planetary systems, all held together by gravity.

**globular cluster:** Old star systems at the edges of spiral galaxies that can contain anywhere from thousands to millions of stars, packed in a close, roughly spherical form and held together by gravity.

**greatest eastern elongation:** When an inner planet (Mercury or Venus) is farthest from the Sun in the western evening sky.

**greatest western elongation:** When an inner planet (Mercury or Venus) is farthest from the Sun in the eastern morning sky.

**light-year:** A unit of distance that uses the distance light travels in one Earth year as its base metric. One light-year is about 9 trillion kilometers (6 trillion miles). Objects in outer space are so far apart that it takes a long time for their light to reach Earth, and so the farther an object is, the farther in the past that observers on Earth are seeing it. As an example, the Andromeda Galaxy (Messier 31) is 2.5 million light-years away, so when observers look at it in the sky, they are seeing it as it appeared 2.5 million years ago.

**magnitude:** The apparent brightness of an object in the sky. Magnitude is measured on a scale where the higher the number, the fainter the object appears. Objects with negative numbers are brighter than those with positive numbers.

**meteor shower:** A celestial event during which a number of meteors can be seen radiating from one point in the sky. Meteor showers occur when Earth, at a particular point in its orbit, crosses a stream of particles left over from a passing comet or asteroid.

**meteor shower peak:** The best time to view a meteor shower, when meteor activity will be at its highest.

**nebula:** A cloud of dust and gas visible in the night sky either as a bright patch or a dark shadow against other luminous matter. To learn more about the different types of nebulae, see page 36.

**open cluster:** A star system that contains anything from a dozen to hundreds of stars, in which the stars are spread out. Open clusters are found near the galactic plane, the plane on which most of a galaxy's mass lies.

**opposition:** When an outer planet (Mars, Jupiter, Saturn, Uranus or Neptune) is opposite the Sun in the sky in right ascension.

**partial lunar eclipse:** A lunar eclipse during which the Earth's shadow partially covers the Moon.

**partial solar eclipse:** A solar eclipse during which the Moon only partially covers the Sun.

**perigee:** When the Moon (or any satellite of Earth) is at its closest in its monthly orbit around Earth.

**perihelion:** For an object orbiting the Sun, the point in its orbit that is nearest to the Sun.

**star:** A luminous ball of plasma held together by its own gravity and fueled initially by the nuclear fusion of hydrogen into helium at its core. Stars come in a vast variety of sizes, luminosities and temperatures, typically ranging from dwarf sizes (that are as little as 10 percent the mass of the Sun) to hypergiants (that can be 100 or more times the mass of the Sun). To learn more about the different types of stars, see pages 10–11.

**summer solstice:** The point during the year that a particular hemisphere is most tilted toward the Sun. This takes place in June in the Northern Hemisphere.

**supernova:** A powerful and luminous explosion that occurs during the last evolutionary stages of a massive star when its brightness increases immensely and it throws off most of its mass; a supernova can also be when a white dwarf is triggered into runaway nuclear fusion by the accretion of material from a binary companion; or by a stellar merger.

**total lunar eclipse:** A lunar eclipse during which the Earth's shadow completely covers the Moon. Such an eclipse can last for hours.

**total solar eclipse:** A solar eclipse during which the Moon covers the entire face of the Sun, revealing the Sun's corona and prominences. Solar eclipses can last from a few seconds to about seven minutes maximum at a specific location.

**variable star:** A star whose apparent magnitude varies over time. The variability might be caused by physical changes to the star itself or by how it rotates or interacts with nearby objects.

**vernal (spring) equinox:** The date in March marking the end of winter and the beginning of spring, when the Sun illuminates the Northern and Southern Hemispheres equally. The lengths of the day and night are equal, and it's one of two instances of the year that the Sun has a declination of 0 degrees

**winter solstice:** The point during the year that a particular hemisphere is most tilted away from the Sun. This takes place in December in the Northern Hemisphere.

**Zenithal Hourly Rate (ZHR):** The rate of meteors a shower would produce per hour under clear, dark skies and with the radiant at the zenith.

## The Greek Alphabet (lowercase)

| | | | | | | | | | |
|---|---|---|---|---|---|---|---|---|---|
| α | alpha | ζ | zeta | λ | lambda | π | pi | φ | phi |
| β | beta | η | eta | μ | mu | ρ | rho | χ | chi |
| γ | gamma | θ | theta | ν | nu | σ | sigma | ψ | psi |
| δ | delta | ι | iota | ξ | xi | τ | tau | ω | omega |
| ε | epsilon | κ | kappa | ο | omicron | υ | upsilon | | |

# Resources

**American Astronomical Society** – aas.org – An organization of professional astronomers seeking to enhance and share humanity's understanding of the Universe

**American Meteor Society** – www.amsmeteors.org – A non-profit scientific organization that informs, encourages and supports research in Meteor Astronomy

**Eclipsewise.com** – www.eclipsewise.com – Astronomer Fred Espenak's site dedicated to predictions and information on eclipses of the Sun and Moon

**European Southern Observatory** – www.eso.org – An intergovernmental research organization for ground-based astronomy that shares recent research, news and images

**European Space Agency (ESA)** – www.esa.int – The main site for the European Space Agency that shares news, research and launches

**Hubblesite.org** – hubblesite.org – NASA's main site for the Hubble Space Telescope, including news and images

**NASA** – www.nasa.gov – An update on all NASA missions, including astronomical events, news and launches

**The Planetary Society** – www.planetary.org – An international, non-profit organization that promotes the exploration of space through education, advocacy and research

**The Royal Astronomical Society of Canada** – www.rasc.ca – Home to Canada's national astronomical association

**Spaceweather.com** – spaceweather.com – An update on space weather, including solar flares, coronal mass ejections and noctilucent cloud forecasts

**Space Weather Prediction Center** – www.swpc.noaa.gov – The National Oceanic and Atmospheric Administration's site on forecasts for space weather, including potential geomagnetic storms

**Solar and Heliospheric Observatory** – soho.nascom.nasa.gov/home.html – A collaborative project between the ESA and NASA to study the Sun and its interactions with Earth from different perspectives

**Solar Dynamics Observatory** – sdo.gsfc.nasa.gov – Satellite observations of the Sun

# Photo Credits

Alan Dyer: 25, 48, 55, 85, 91, 107, 113
Alan Dyer/Stocktrek Images/Alamy Stock Photo: 7
Debra Ceravolo: 5, 14, 37 (top), 49, 77
Elliot Severn: 9
ESA/Hubble and NASA: 87, 95
Fred Espenak: 27, 28, 29 (top and bottom)
Jay Anderson: 61
Malcolm Park: 15, 26, 36, 42, 51, 67, 72, 75, 83, 89, 90
Michael Watson: 20–21
NASA: 34, 35, 53
NASA/Bill Dunford: 18
NASA, ESA and the Hubble Heritage Team (STScI/AURA): 65, 93, 101, 111
NASA, ESA and the Hubble SM4 ERO Team: 99
NASA and the Hubble Heritage Team (STScI/AURA): 69
NASA/Johns Hopkins University Applied Physics Laboratory/Carnegie Institution of Washington: 32
NASA/JPL-Caltech: 33
NASA/SDO: 24 (bottom)
Nicole Mortillaro: 73, 103
Notanee Bourassa: 23, 31
Ron Brecher: 47, 71, 81, 105
Sean Walker: 24 (top), 37 (bottom), 39 (top)
Shelley Jackson: 57, 63
Stefanie Harron: 59
Trevor Jones: 45

**Front cover:** Kerry-Ann Lecky Hepburn